教育中国·畅销精品系列

教育部高等学校材料类专业教学指导委员会规划教材

中国石油和化学工业优秀教材奖

ENGLISH FOR POLYMER MATERIALS AND ENGINEERING

高分子材料工程专业英语
第三版

大学英语专业阅读教材编审委员会　组织编写

曹同玉　冯连芳　张菊华　主编

化学工业出版社

·北　京·

内容简介

本书是"大学英语专业阅读教材编审委员会"组织编写的高等学校专业英语系列教材之一。全书共34课，包括高分子化学、高分子物理、聚合反应工程、聚合物性能、成型加工及应用以及高分子材料的实验、研制与生产等多方面的内容。每课均由课文、重点词汇（单词、音标及解释）、词组、课文注释、练习、阅读材料等部分组成。本书是数字融合教材，配套有原文朗读、教学课件等资源，供读者选择使用。

本书主要作为全国各高等院校高分子材料与工程及相关材料专业的英语教材，也可作为从事高分子合成、成型加工、研制及应用工作的科技人员、教师及研究生提高专业英语水平的参考用书。

图书在版编目（CIP）数据

高分子材料工程专业英语/曹同玉，冯连芳，张菊华主编. —3版. —北京：化学工业出版社，2022.7（2024.7重印）
ISBN 978-7-122-41210-2

Ⅰ.①高… Ⅱ.①曹… ②冯… ③张… Ⅲ.①高分子材料-英语-高等学校-教材 Ⅳ.①TB324

中国版本图书馆CIP数据核字（2022）第060604号

责任编辑：王　婧　杨　菁
责任校对：王　静
装帧设计：李子姮

出版发行：化学工业出版社
　　　　　（北京市东城区青年湖南街13号　邮政编码100011）
印　　装：河北延风印务有限公司
787mm×1092mm　1/16　印张12$\frac{3}{4}$　字数344千字
2024年7月北京第3版第4次印刷

购书咨询：010-64518888
售后服务：010-64518899
网　　址：http://www.cip.com.cn

凡购买本书，如有缺损质量问题，本社销售中心负责调换。

定　　价：39.00元　　　　　　　　　　　　　　　版权所有　违者必究

全国部分高校化工类及相关专业
大学英语专业阅读教材编审委员会

主 任 委 员　朱炳辰　华东理工大学

副主任委员　吴祥芝　北京化工大学
　　　　　　　钟　理　华南理工大学
　　　　　　　欧阳庆　四川大学
　　　　　　　贺高红　大连理工大学

委　　　员　赵学明　天津大学
　　　　　　　张宏建　浙江大学
　　　　　　　王延儒　南京化工大学
　　　　　　　徐以撒　江苏石油化工学院
　　　　　　　魏新利　郑州大学
　　　　　　　王　雷　辽宁石油化工大学
　　　　　　　胡惟孝　浙江工业大学
　　　　　　　吕廷海　北京石油化工学院
　　　　　　　陈建义　中国石油大学
　　　　　　　胡　鸣　华东理工大学

秘　　　书　何仁龙　华东理工大学教务处

第三版前言

20世纪90年代后期,由化工部"大学英语专业阅读教材编审委员会"组织多所化工院校成立了"高等学校化工专业英语系列教材编审委员会",该委员会由华东理工大学、北京化工大学、华南理工大学、四川大学、大连理工大学、天津大学、浙江大学、南京化工大学等十四所化工专业实力较强的高等院校组成,华东理工大学资深教授朱炳辰先生任主任委员。在"大学英语专业阅读教材编审委员会"的领导下,在化学工业出版社的主持下,"高等学校化工专业英语系列教材编审委员会"按专业分头编写了一套高等学校化工专业英语系列教材。这套书其中一册为高分子专业英语,即《高分子材料工程专业英语》,其第一版是由天津大学(曹同玉)、浙江大学(冯连芳)及四川大学(张菊华等)联合编写的,1998年完稿,1999出版发行。

在第一版出版以后,本书受到了广大读者的关注,很快成为一本畅销书,重印多次后,于2011年发行了第二版。

自本教材第二版出版发行至今已有十二年之久,在这期间又重印了10余次,鉴于采用此教材的各校师生在教学和阅读过程中发现了在此书中所存在的缺陷和错误需要修改,同时,作为一本高等学校教材也应当不断提炼、不断更新、不断充实、不断提高,以期获得更好的教学效果。因此,受化学工业出版社的委托,在向多所高等院校和高分子领域的学者们进行了调查研究的基础上,由曹同玉执笔,对本书第二版进行了全面审订、勘误、修改、删节、增补和重新编排,进行了仔细的文字推敲,并充实了聚合物加工方面的内容;另外,还增加了对互联网的应用,引入了在线教学学习资源——课文朗读和PPT教学课件,课文朗读可以增强学生的英语口语能力,而PPT教学课件可以供教师编辑在课堂上拟深入讲解的内容,能为教学提供方便。课文朗读和教学课件均可以通过扫描书上相应位置的二维码来获取。全书共分34课,包括高分子化学、高分子物理、聚合反应工程、聚合物性能、加工成型与应用以及高分子材料的实验、研发及生产等多方面的内容。每一课均由课文、重点词汇(单词、音标及解释)、词组、课文注释、练习、阅读材料等多方面的内容组成。每一课的课文和阅读材料均为彼此独立的短文,分别取材于不同国家、不同作者的50余种英文高分子书籍、期刊、会议论文等英文原文资料,集不同语言风格为一体,具有词汇量大及科技英语语法覆盖面广等特点。本书主要作为全国各高等院校高分子专业英语教材,也可以作为从事聚合物合成、成型加工、研发及应用工作的科技工作者及研究生等人员提高专业英语水平的参考书。

本书分34课(UNIT),其中UNIT 1～UNIT 12由曹同玉、袁才登编写,UNIT 13～UNIT 22由冯连芳编写,UNIT 23、UNIT 28及UNIT 34由李瑞海编写,UNIT 24、UNIT 25及UNIT 33由雷勇编写,UNIT 26、UNIT 31及UNIT 32由孙树东编写,UNIT 27、UNIT 29及UNIT 30由胡泽容编写,PART C部分由张菊华和欧阳庆主审,全书由曹同玉统审和定稿。

在本书第三版编写过程中,承蒙得到了袁才登、赵万里、谢洪云、何跃华、李淑荣、曹汇川等同志多方面的帮助,尤其是陈奕安同志在课文勘误、引入互联网在线教学与学习方面给予了大力支持,编者深表谢意。

限于编者水平,再版后书中难免还会有疏漏和不尽人意之处,恳请各位老师、各位同学以及其他各行业的相关读者批评指正。

编者
2022.2

第二版前言

《高分子材料工程专业英语》第一版是由"大学英语专业阅读教材编审委员会"于1998年组织编写的高等学校化工专业英语系列教材之一。该教材编审委员会是由华东理工大学、北京化工大学、华南理工大学、四川大学、大连理工大学、天津大学、浙江大学、南京化工大学等十四所化工专业实力较强的高等院校组成的,由华东理工大学资深教授朱炳辰先生任主任委员。本书第一版是在大学英语专业阅读教材编审委员会的领导下,在化学工业出版社的主持下,由天津大学(曹同玉)、浙江大学(冯连芳)、四川大学(张菊华等)联合编写的。

自从1999年第一版出版以来,本书受到了广大读者的关注,已被全国许多高等院校和部分中等专业学校的高分子材料专业、高分子化工专业、材料科学与工程专业等专业用做专业英语教材;同时也有不少在高分子科学与技术领域里工作的工程技术人员、研究人员、准备晋升职称的人员、准备出国留学人员、在校研究生等把此书作为提高高分子专业英语水平的基本阅读材料,受到了广大读者的青睐,给予了较高的评价,同时也提出了许多宝贵意见和建议,使编者受益匪浅,在此向广大读者深表谢意。

本书第一版出版至今已有十多年之久,已重印10余次,鉴于采用此教材的各校师生在专业英语教学过程中发现了书中存在的一些缺陷,需要修改;同时,作为一本高等学校教材也应当不断提炼,不断更新,不断提高,以期获得更好的教学效果;再者,近年来高分子科学又有了新的发展,在教材中应当有所反映,有所充实,因此,化学工业出版社决定再版此书。受化学工业出版社的委托,在向多所院校相关师生进行了调查研究的基础上,由曹同玉执笔,对本书第一版进行了全面的重新审订、勘误、修改、删节、增补和编排,认真地进行了文字推敲;同时还引入了在高分子领域里新近出现的活性自由基聚合等新章节,并充实了高分子物理方面的相关内容,可望《高分子材料工程专业英语》第二版能更好地满足广大读者对高分子专业英语教材越来越高的要求。

在本书第二版的修订和编写过程中,得到了袁才登、瞿雄伟、肖继君、郭睿威、董岸杰、赵万里等各位同仁的大力帮助,并提出了许多宝贵意见和建议,编者表示诚挚的谢意。

限于编者水平,再版后的书中一定还会有不尽如人意之处,恳请各位老师、各位同学及其他读者批评指正。

编者
2011.1

第一版前言

组织编审出版系列的专业英语教材，是许多院校多年来共同的愿望。在高等教育面向 21 世纪的改革中，学生基本素质和实际工作能力的培养受到了空前重视。对非英语专业的学生而言，英语水平和能力的培养不仅是文化素质的重要部分，在很大程度上也是能力的补充和延伸。在此背景下，教育部（原国家教委）几次组织会议研究加强外语教学问题，制订相关规范，使外语教学更加受到重视。教材是教学的基本要素之一，与基础英语相比，专业英语教学的教材问题此时显得尤为突出。

国家主管部门的重视和广大院校的呼吁引起了化学工业出版社的关注，他们及时地与原化工部教育主管部门和全国化工类专业教学指导委员会请示协商后，组织全国十余所院校成立了大学英语专业阅读教材编委会。在经过必要的调研后，根据学校需求，编委会优先从各校教学（交流）讲义中确定选题，同时组织力量开展编审工作。本套教材涉及的专业主要包括化学工程与工艺、石油化工、机械工程、信息工程、生产过程自动化、应用化学及精细化工、生化工程、环境工程、制药工程、材料科学与工程、化工商贸等。

于 20 世纪 70 年代末期，全国各化工院校高分子材料与高分子化工专业都相继开设了"高分子专业英语"课。这门课为完成由基础英语向专业英语的过渡，提高学生阅读高分子专业文献资料的能力发挥了重要作用。但是，对于"高分子专业英语"这门课，不同学校，即使同一学校的不同教师，甚至同一教师在不同年份，所用教材、教学内容、教学要求和教学方法都不尽相同，给人们的印象是这门课的随意性很大，没有一定可遵循的规范，教学质量也因地、因人、因时而异，这很不利于学生专业英语能力的培养。如果能编写一本"高分子专业英语"全国统编教材，各学校都按照这本教材的内容和要求进行规范的教学，对于克服目前"高分子专业英语"课教学的混乱状态，无疑会起到至关重要的作用。

这本《高分子材料工程专业英语》即是根据"全国部分高校化工类及相关专业大学英语专业阅读教材编审委员会"的要求和安排编写的。全书共分 32 课，包括高分子化学、高分子物理、聚合反应工程、聚合物性能、成型加工及应用，以及高分子材料的实验、研制与生产等多方面的内容。每课均由课文、重点词汇（单词、音标及解释）、词组、课文注释、练习、阅读材料等部分构成。每课的课文和阅读材料均为彼此独立的短文，取材于不同国家、不同作者的 48 种英文高分子专业书籍、会议论文集、期刊、专利等英文原文资料，集不同语言风格为一体，具有词汇量大及科技英语语法覆盖面广等特点。本书主要作为全国各高等院校高分子专业英语教材，也可以作为从事高分子合成、成型加工、研制及应用工作的科技人员、教师及研究生提高专业英语水平的参考书。

本书分工情况为：Unit1～Unit10 由曹同玉、袁才登编写，Unit11～Unit20 由冯连芳编写，Unit21、Unit27 及 Unit28 由李瑞海编写，Unit22、Unit23 及 Unit26 由雷勇编写，Unit24、Unit31及Unit32 由孙树东编写，Unit25、Unit29 及 Unit30 由胡泽容编写。全书由曹同玉、冯连芳主编，由张菊华、欧阳庆主审。在编写过程中承蒙方道斌、姚兆玲、刘德华、陈锦言、赵勇等同志进行了审阅与校核，提出了许多宝贵意见，并给予了多方面的帮助，在此深表谢意。限于作者水平，书中一定会有不少错误，望读者批评指正。

编者
1998.8

目 录

PART A Polymer Chemistry and Physics ········ 1
- UNIT 1   What Are Polymers? ········ 1
- UNIT 2   Chain Polymerization ········ 7
- UNIT 3   Step-Growth Polymerization ········ 13
- UNIT 4   Ionic Polymerization ········ 19
- UNIT 5   Introduction to Living Radical Polymerization ········ 25
- UNIT 6   Molecular Weight and Its Distributions of Polymers ········ 30
- UNIT 7   Polymer Solution ········ 36
- UNIT 8   Morphology of Solid Polymers ········ 42
- UNIT 9   Structure and Properties of Polymers ········ 48
- UNIT 10 Glass Transition Temperature ········ 53
- UNIT 11 Functional Polymers ········ 58
- UNIT 12 Preparations of Amino Resins in Laboratory ········ 64

PART B Polymerization Reaction Engineering ········ 71
- UNIT 13 Reactor Types ········ 71
- UNIT 14 Bulk Polymerization ········ 77
- UNIT 15 General Description of VC Suspension Polymerization Process ········ 81
- UNIT 16 Styrene-Butadiene Copolymer ········ 86
- UNIT 17 Heat Transfer Process ········ 93
- UNIT 18 Copolymer Composition Distributions Affected by Micromixing ········ 98
- UNIT 19 Introduction to Modelling of Polymerization Kinetics ········ 104
- UNIT 20 Polymerization Process Instrumentation ········ 110
- UNIT 21 Reactor Scale-up ········ 116
- UNIT 22 UNIPOL Process for Polyethylene ········ 121

PART C Processing, Properties and Applications of Polymer Material ········ 127
- UNIT 23 Polymer Processing ········ 127
- UNIT 24 Mechanical Properties of Polymers ········ 133
- UNIT 25 Thermal Properties of Polymers ········ 137
- UNIT 26 Polymer Melts ········ 141
- UNIT 27 Processing and Fabrication of Thermoplastics ········ 145
- UNIT 28 General Aspects of Polymer Degradation ········ 151
- UNIT 29 Synthetic Plastics ········ 156
- UNIT 30 Synthetic Rubber ········ 162
- UNIT 31 Structure of Fiber-forming Polymers ········ 169
- UNIT 32 Matching Adhesive to Adherend ········ 174
- UNIT 33 Processing of Thermosets ········ 178
- UNIT 34 Fillers for Polymers ········ 182

APPENDIXES 聚合物的命名法 ········ **188**

PART A

Polymer Chemistry and Physics

UNIT 1 What Are Polymers?

What are polymers? For one thing, they are complex and giant molecules and are different from low molecular weight compounds like, say, common salt. To contrast the difference, the molecular weight of common salt is only 58.5, while that of a polymer can be as high as several hundred thousands, even more than thousand thousands. These big molecules or 'macro-molecules' are made up of much smaller molecules. The small molecules, which combine to form a big molecule, can be of one or more chemical compounds. To illustrate, imagine that a set of rings has the same size and is made of the same material. When these rings are interlinked, the chain formed can be considered as representing a polymer from molecules of the same compound. Alternatively, individual rings could be of different sizes and materials, and interlinked to represent a polymer from molecules of different compounds.

This interlinking of many units has given the polymer its name, *poly* meaning 'many' and *mer* meaning 'part' (in Greek). As an example, a gaseous compound called butadiene, with a molecular weight of 54, combines nearly 4000 times and gives a polymer known as polybutadiene (a synthetic rubber) with about 200000 molecular weight. The low molecular weight compounds from which the polymers form are known as monomers. The picture is simply as follows:

$$\text{butadiene} + \text{butadiene} + \cdots + \text{butadiene} \longrightarrow \text{polybutadiene}$$
$$(4000 \text{ times})$$

One can thus see how a substance (monomer) with as small a molecular weight as 54 grows to become a giant molecule (polymer) of $(54 \times 4000 \approx)$ 200000 molecular weight. It is essentially the 'giantness' of the size of the polymer molecule that makes its behavior different from that of a commonly known chemical compound such as benzene.① Solid benzene, for instance, melts to become liquid benzene at 5.5℃ and, on further heating, boils into gaseous benzene. As against this well-defined behavior of a simple chemical compound, a polymer like polyethylene does not melt sharply at one particular temperature into clean liquid. Instead, it becomes increasingly softer and, ultimately, turns into a very viscous, tacky molten mass. Further heating of this hot, viscous, molten polymer does convert it into various gases but it is no longer polyethylene (Fig. 1.1).

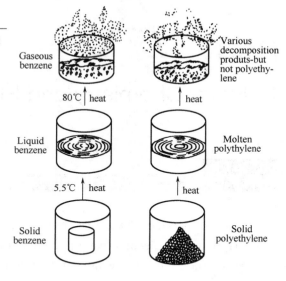

Fig. 1.1

Difference in behavior on heating of a low molecular weight compound (benzene) and a polymer (polyethylene)

Another striking difference with respect to the behavior of a polymer and that of a low molecular weight compound concerns the dissolution process. Let us take, for example, sodium chloride and add it slowly to a fixed quantity of water. The salt, which represents a low molecular weight compound, dissolves in water up to a point (called saturation point) but, thereafter, any further quantity added does not go into solution but settles at the bottom and just remains there as solid. The viscosity of the saturated salt solution is not very much different from that of water. But if we take a polymer instead, say, polyvinyl alcohol, and add it to a fixed quantity of water, the polymer does not go into solution immediately. The globules of polyvinyl alcohol firstly absorb water, swell and get distorted in shape and after a long time go into solution.② Also, we can add a very large quantity of the polymer to the same quantity of water without the saturation point ever being reached. As more and more quantity of polymer is added to water, the time taken for the dissolution of the polymer obviously increases and the mix ultimately assumes a soft, dough-like consistency. Another peculiarity is that, in water, polyvinyl alcohol never retains its original powdery nature as the excess sodium chloride does in a saturated salt solution.③ In conclusion, we can say that (1) the long time taken by polyvinyl alcohol for dissolution, (2) the absence of a saturation point, and (3) the increase in the viscosity are all characteristics of a typical polymer being dissolved in a solvent and these characteristics are attributed mainly to the large molecular size of the polymer. The behavior of a low molecular weight compound and that of a polymer on dissolution are illustrated in Fig. 1.2.

——Gowariker V R, Viswanathan N V, Sreedhar J. Polymer Science. New York: John Wiley & Sons, 1986.6

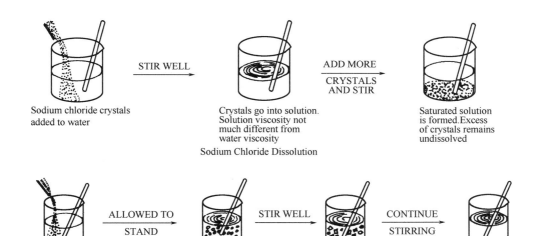

Fig. 1.2

Difference in solubility behaviour of a low molecular weight compound (sodium chloride) and a polymer (polyvinyl alcohol)

Words and Expressions

polymer	['pɔlimə]	n.	聚合物[体]，高聚物
common salt			食盐
macromolecule	[mækrə'mɔlikju:l]	n.	大分子，高分子
imagine	[i'mædʒin]	v.	想象，推测
interlink	[intə'liŋk]	v.	把…相互连接起来
		n.	连接
butadiene	[bju:tə'daii:n]	n.	丁二烯
monomer	['mɔnəmə]	n.	单体
synthetic	[sin'θetik]	a.	合成的
behavior	[bi'heivjə]	n.	性能，行为
polyethylene	[pɔli'eθili:n]	n.	聚乙烯
viscous	['vəskəs]	a.	黏稠的
tacky	['tæki]	a.	（表面）发黏的
		n.	粘连[搭]性
dissolution	[disə'lu:ʃən]	n.	溶解
dissolve	[di'sɔlv]	v.	使…溶解
saturation	[sætʃə'reiʃən]	n.	饱和
settle	['setl]	v.	沉淀[降]，澄清
viscosity	[vis'kɔsiti]	n.	黏度[性]
polyvinyl alcohol			聚乙烯醇
globule	['glɔbju:l]	n.	小球，液滴，颗粒
swell	[swel]	v.; n.	溶胀
distort	[dis'tɔ:t]	v.	使…变形，扭曲
dough	[dəu]	n.	（生）面团，揉好的面
consistency	[kɔn'sistənsi]	n.	稠度，黏稠性
powdery	['paudəri]	a.	粉状的
solvent	['sɔlvent]	n.	溶剂
peculiarity	[pikju:li'æriti]	n.	特性
crystal	['kristl]	n.	晶体，结晶
fragment	['fræɡmənt]	n.	碎屑，碎片

Phrases

for one thing　首先
as an example　例如，举例来说
as against　和…比起来，和…相对照
convert…into…　把…转变［化］成…

with respect to　关于，就…而论
a quantity of…　大量，一些
in conclusion　总之，最后
be attributed to…　归因于，认为是…的结果

Notes

① "It is essentially the 'giantness' of the size of the polymer molecule that makes its behavior different from that of a commonly known chemical compound such as benzene." 此处 It is… that…为强调语气结构。the 'giantness' of the size of the polymer molecule 为被强调的句子的主语，可译为"巨大的聚合物分子尺寸"。句子的谓语是"makes"。"its behavior"是"makes"的宾语，可译为"聚合物的性能"。"different from…"为宾语补足语，其后的"that"代替前面已出现过的"behavior"，以避免重复。该句译文为："实质上，正是由于聚合物的巨大的分子尺寸才使其性能不同于像苯这样的一般化合物（的性能）。"

② "The globules of polyvinyl alcohol firstly absorb water, swell and get distorted in shape and after a long time go into solution." 是一个简单句，"the globules"为主语，作"球粒"或"颗粒"解。本句有四个并列的谓语，即 absorb、swell、get distorted 及 go into。"get distorted in shape"为"get + 过去分词"表示的被动语态。全句的译文是："聚乙烯醇颗粒首先吸水溶胀，发生形变，经过很长的时间以后，（聚乙烯醇分子）进入到溶液中。"

③ "Another peculiarity is that, in water, polyvinyl alcohol never retains its original powdery nature as the excess sodium chloride does in a saturated salt solution." 从属连接词 that 所引导的从句是表语从句，从句的宾语"its original powdery nature"可译成"其初始的粉末状态"。最后是 as 引导的方式方法状语从句，从句中的"does"用于替代前面出现的动词"retains"。全句译文为："另一个特点是，在水中聚乙烯醇不会像过量的氯化钠在饱和盐溶液中那样能保持其初始的粉末状态。"

Exercises

1. *Translate the following into Chinese*

   Not all polymers are built up from bonding together a single kind of repeating unit. At the other extreme, protein molecules are polyamides in which n amino acid repeat units are bonded together. Although we might still call n the degree of polymerization in this case, it is less useful, since an amino acid unit might be any one of some 20-odd molecules that are found in proteins. In this case the molecular weight itself, rather than the degree of polymerization, is generally used to describe the molecule. When the actual content of individual amino acids is known, it is their sequence that is of special interest to biochemists and molecular biologists.

2. *Give a definition for each following word*
   （1）molecule
   （2）monomer
   （3）polymer

3. *Put the following words into Chinese*
   structure　data　equation　pressure　liquid　laboratory　solid
   molecule　temperature　measurement　compound　electrical

4. *Put the following words into English*
   科学　技术　化学　物理　气体　原子　性质　试验　增加　减少　混合物

Reading Materials

Structure of Polymer Chains

In many cases polymer chains are linear. In evaluating both the degree of polymerization and the extended chain length, we assume that the chain has only two ends. While linear polymers are important, they are not the only type of molecules possible. Branched and cross-linked molecules are also important. When we speak of a branched polymer, we refer to the presence of additional polymeric chains issuing from the backbone of a linear molecule. Substituent groups such as methyl or phenyl groups on the repeat units are not considered branches. Branching is generally introduced into a molecule by intentionally adding some monomer with the capability of serving as a branch. Let us consider the formation of a polyester. The presence of difunctional acids and difunctional alcohols allows the polymer chain to grow. These difunctional molecules are incorporated into the chain with ester linkages at both ends of each. Trifunctional acids or alcohols, on the other hand, produce a linear molecule by reacting two of their functional groups. If the third reacts and the resulting chain continues to grow, a branch has been introduced into the original chain. Adventitious branching sometimes occurs as a result of an atom being abstracted from the original linear molecule, with chain growth occurring from the resulting active site. Molecules with this kind of accidental branching are generally still called linear, although the presence of significant branching has profound effects on some properties of the polymer, most notably the tendency to undergo crystallization.

The amount of branching introduced into a polymer is an additional variable that must be specified for the molecule to be fully characterized. When only a slight degree of branching is present, the concentration of junction points is sufficiently low that these may be simply related to the number of chain ends. For example, two separate linear molecules have a total of four ends. If the end of one of these linear molecules attaches itself to the middle of the other to form a "T", the resulting molecule has three ends. It is easy to generalize this result. If a molecule has ν branches, it has $\nu + 2$ chain ends if the branching is relatively low. Branched molecules are sometimes described as either combs or stars. In the former, branch chains emanate from along the length of a common backbone; in the latter, all branches radiate from a central junction.

If the concentration of junction points is high enough, even branches will contain branches. Eventually a point is reached at which the amount of branching is so extensive that the polymer molecule becomes a giant three dimensional network. When this condition is achieved, the molecule is said to be cross-linked. In this case, an entire macroscopic object may be considered to consist of essentially one molecule. The forces which give cohesiveness to such a body are covalent bonds, not intermolecular forces. Accordingly, the mechanical behavior of cross-linked bodies is much different from those without cross-linking.

Just as it is not necessary for polymer chains to be linear, it is also not necessary for all repeat units to be the same. We have already mentioned molecules like proteins where a wide variety of different repeat units are present. Among synthetic polymers, those in which a single kind of repeat unit are involved are called homopolymers, and those containing more than one kind of repeat unit are copolymers. Note that these definitions are based on the repeat unit, not the monomer. An ordinary polyester is not a copolymer, even though two different monomers, acids and alcohols, are its monomers. By contrast, copolymers result when different monomers bond together in the same way to produce a chain in which each kind of monomer retains its respective substituents in the polymer molecule. The unmodified term *copolymer* is generally used to designate the case where two different repeat units are involved. Where three kinds of repeat units are present, the system is called a terpolymer; where there are more than three, the system is called a multicomponent copolymer.

The moment we admit the possibility of having more than one kind of repeat unit, we require additional variables to describe the polymer. First, we must know how many kinds of repeat units are present and what they are. This is analogous to knowing what components are present in a solution, although the similarity ends there, since the repeat units in a polymer are bonded together and not merely mixed. To describe the copolymer quantitatively, the relative amounts of the different kinds of repeat units must be specified. Thus the empirical formula of a copolymer may be written $A_x B_y$, where A and B signify the individual repeat units and x and y indicate the relative number of each. From a knowledge of the molecular weight of the polymer, the molecular weights of A and B, and the values of x and y, it is possible to calculate the number of each kind of monomer unit in the copolymer. The sum of these values gives the degree of polymerization of the copolymer. Note that we generally do not call n_A and n_B the degrees of polymerization of the individual units. The inadvisability of the latter will become evident presently.

——Hiemenz P C. Polymer Chemistry. New York: Marcel Dekker, 1984. 9

Words and Expressions

linear polymer			线型聚合物
branched polymer			支链聚合物
homopolymer	[hɔməˈpɔlimə]	n.	均聚物
backbone	[ˈbækbəun]	n.	主链
polyester	[pɔliˈestə]	n.	聚酯
difunctional	[diˈfʌŋkʃənl]	a.	二［双］官能度的
crystallization	[kristəlaiˈzeiʃən]	n.	结晶（作用）
emanate	[ˈeməneit]	v.	源于，发源，发出，放射（出）
cohesiveness	[kəuˈhiːsivnis]	n.	内聚性［力］，黏结性
terpolymer	[təːˈpɔlimə]	n.	三元共聚物
inadvisability	[inədvaizəˈbiliti]	n.	不合理（性），不适当

UNIT 2 Chain Polymerization

Many olefinic and vinyl unsaturated compounds are able to form chain-like macromolecules through elimination of the double bond, a phenomenon first recognized by Staudinger. Diolefins polymerize in the same manner, however, only one of the two double bonds is eliminated. Such reactions occur through the initial addition of a monomer molecule to an initiator radical or an initiator ion, by which the active state is transferred from the initiator to the added monomer. ① In the same way, by means of a chain reaction, one monomer molecule after the other is added (2000~20000 monomers per second) until the active state is terminated through a different type of reaction. The polymerization is a chain reaction in two ways: because of the reaction kinetics and because as a reaction product one obtains a chain molecule. The length of the chain molecule is proportional to the kinetic chain length.

One can summarize the process as follows (R· is equal to the initiator radical):

$$R\cdot + CH_2{=}CH(Cl) + CH_2{=}CH(Cl) + CH_2{=}CH(Cl) + \cdots \longrightarrow$$
$$R{-}CH_2{-}CH(Cl){-}CH_2{-}CH(Cl){-}CH_2{-}CH(Cl)\sim\sim$$

One thus obtains polyvinylchloride from vinylchloride, or polystyrene from styrene, or polyethylene from ethylene, etc.

The length of the chain molecules, measured by means of the degree of polymerization, can be varied over a large range through selection of suitable reaction conditions. Usually, with commercially prepared and utilized polymers, the degree of polymerization lies in the range of 1000 to 5000, but in many cases it can be below 500 and over 10000. This should not be interpreted to mean that all molecules of a certain polymeric material consist of 500, or 1000, or 5000 monomer units. In almost all cases, the polymeric material consists of a mixture of polymer molecules of different degrees of polymerization.

Polymerization, a chain reaction, occurs according to the same mechanism as the well-known chlorine-hydrogen reaction and the decomposition of phosgene.

The initiation reaction, which is the activation process of the double bond, can be brought about by heating, irradiation, ultrasonics, or initiators. The initiation of the chain reaction can be observed most clearly with radical or ionic initiators. ② These are energy-rich compounds which can add suitable unsaturated compounds (monomers) and maintain the activated radical or ionic state so that further monomer molecules can be added in the same manner. ③ For the individual steps of the growth reaction one needs only a relatively small activation energy and therefore through a single activation step (the actual initiation reaction) a large number of olefin molecules are converted, as is implied by the term "chain reaction". ④ Because very small amounts of the initiator bring about the formation of a large amount of polymeric material (1 : 1000 to 1 : 10000), it is possible to regard polymerization from a superficial point of view as a catalytic reaction. For this reason, the initiators used in polymerization reactions are often designated as polymerization catalysts, even though, in the strictest sense, they are not true catalysts because the polymerization initiator enters into the reaction as a real partner and can be found chemically bound in the reaction product, i.e., the polymer. In addition to the ionic and radical initiators there are now metal complex initiators (which can be obtained, for example, by the reaction of titanium tetrachloride or titanium trichloride with aluminum alkyls), which play an important role in polymerization reactions (Ziegler catalysts). The mechanism of their catalytic action is not yet completely clear.

——Vollmert B. Polymer Chemistry. Berlin: Springer-Verlag, 1973. 40

Words and Expressions

olefinic	[əuləˈfinik]	a.	烯烃的
vinyl	[ˈvainil]	n.; a.	乙烯基（的）
unsaturated	[ʌnˈsætʃəreitid]	a.	不饱和的
eliminate	[iˈlimineit]	v.	消除，打开，除去
double bond		n.	双键
diolefin	[daiˈəuləfin]	n.	二烯烃
transfer	[ˈtræsfə:]	v.	（链）转移，（热）传递
initiator	[iˈniʃieitə]	n.	引发剂
radical	[ˈrædikəl]	n.	自由基
chain reaction			连锁反应
terminate	[ˈtə:mineit]	v.	（链）终止
kinetic chain length			动力学链长
polyvinylchloride	[pɔliˈvainilˈklɔraid]	n.	聚氯乙烯
polystyrene	[pɔliˈstaiərin]	n.	聚苯乙烯
degree of polymerization			聚合度
polymeric	[pɔliˈmerik]	a.	聚合（物）的
mechanism	[ˈmekənizəm]	n.	机理 [制]
chlorine	[ˈklɔ:ri:n]	n.	氯（气）
hydrogen	[ˈhaidridʒən]	n.	氢（气）
decomposition	[di:kɔmpəˈziʃən]	n.	分解
phosgene	[ˈfɔzdʒi:n]	n.	光气，碳酰氯
initiation	[iniʃiˈeiʃən]	n.	（链）引发
activation	[æktiˈveiʃən]	n.	活化（作用）
irradiation	[ireidiˈeiʃən]	n.	照射，辐照
ultrasonic	[ʌltrəˈsɔnik]	n.	超声波
catalyst	[ˈkætəlist]	n.	催化剂，触媒
ionic	[aiˈɔnik]	a.	离子的
complex	[ˈkɔmpleks]	n.	络合物
titanium tetrachloride			四氯化钛
titanium trichloride			三氯化钛
aluminum alkyl			烷基铝

Phrases

by means of... 借助于…
one...after the other 一个接一个…
be proportional to... 和…成正比
over a large range 在很大的范围内
lie in... 处于，落在，在于

bring about 引起，产生，导致
energy-rich 高能（级）的
from a superficial point of view 从表面上看
in the strictest sense 严格地讲 [说]
play an important role in... 在…方面起重要作用

Notes

① "Such reactions occur through the initial addition of a monomer molecule to an initiator radical or an initiator ion, by which the active state is transferred from the initiator to the added monomer." 主语 "such reaction"是指上面所提到的打开单体双键而形成聚合物的反应。谓语"occur"后面为由"through"引导的介词短语作方式状语,可译为"这些反应是通过…而进行的"。最后为非限定性的定语从句,不是说明某个名词,而是说明前面整个句子。"by which"在从句中作方式状语。"active state"可译作"活性中心"。全句译文:"这样的反应是通过单体分子首先加成到引发剂自由基或引发剂离子上而进行的,靠这些反应活性中心由引发剂转移到被加成的单体上。"

② "The initiation of the chain reaction can be observed most clearly with radical or ionic initiators." 该句是一个简单句,句末介词短语"with radical or ionic initiators"是主语"the initiation"的定语,为了使句子匀称,避免头重脚轻,后置定语没有紧接它要说明的主语,而是被其他成分如谓语等分隔开,这在英语语法中称作分割现象,被分割的定语可以是介词短语,也可以是定语从句。该句是一个起后置定语作用的介词短语被分割的例句,可译作:"用自由基型引发剂或离子型引发剂引发连锁反应可以很清楚地进行观察。"

③ "These are energy-rich compounds which can add suitable unsaturated compounds (monomers) and maintain the activated radical, or ionic, state so that further monomer molecules can be added in the same manner." 主句的表语"energy-rich compounds"意指"高能态化合物",紧跟其后的句子为以"which"引导的定语从句。最后为由"so that"引导的结果从句。全句译文为:"这些(化合物)是高能态化合物,它们可以加成不饱和化合物(单体),并且(在完成一步加成以后仍然)可以保持自由基活性中心或离子活性中心,致使单体分子可以用同样的方式进一步加成。"

④ "For the individual steps of the growth reaction one needs only a relatively small activation energy and therefore through a single activation step (the actual initiation reaction) a large number of olefin molecules are converted, as is implied by the term 'chain reaction.'" 句中主语"one"用以代表前面的复数名词"steps","one"是单数,意指在"steps"中的一步。句子的最后为as引导的定语从句,用以说明前面整个句子,"as"在从句中作主语。全句译文为:"对于链增长反应的诸多步骤来说,每一步仅需要相当少的活化能,因此,通过一步简单的活化反应(即引发反应)即可将许多烯类单体分子转化(成聚合物),这正如连锁反应这个术语的内涵那样。"

Exercises

1. *Please fill in the correct answers into the blanks in the following passage*

Another striking difference with respect to the behavior of a polymer and _____ of a low molecular weight compound concerns the dissolution process. Let us take, for example, sodium chloride and add it slowly to a fixed _____ of water. The salt, which represents a _____ molecular weight compound, dissolves in water up to a point (called _____ point) but, thereafter, any further quantity added does not go into solution but settles at the _____ and just remains there as solid. The viscosity of the saturated salt solution is not very _____ different from that of water. But if we take a polymer instead, say, polyvinyl alcohol, and add it to a fixed quantity of water, the polymer does not go into solution immediately. The globules of polyvinyl alcohol first _____ water, swell and get distorted in shape and after a long time go into solution.

2. *Translate the following into English*

乙烯分子带有一个双键,为一种烯烃,它可以通过连锁聚合大量地制造聚乙烯,目前,聚乙烯已经广泛地应用于许多技术领域和人们的日常生活中,成为一种不可缺少的材料。

3. *Put the following words into Chinese*

macromolecule   tacky   settle   behavior   molten polymer   distort   viscous   butadiene   synthetic   globule   powdery   fragment

4. *Put the following words into English*

氯化钠   黏度   吸收   溶胀   单体   苯   分子量   化合物   溶液   形状   低分子化合物   高分子化合物

Reading Materials

Overall Kinetics of Chain Polymerization

Radical chain polymerization is a chain reaction consisting of a sequence of three steps—*initiation, propagation and termination*. The initiation step is considered to involve two reactions. The first is the production of free radicals by any one of a number of reactions. The usual case is the homolytic dissociation of an initiator or catalyst species I to yield a pair of radicals R·

$$I \xrightarrow{k_d} 2R \cdot \quad (2.1)$$

where k_d is the rate constant for the catalyst dissociation. The second part of the initiation involves the addition of this radical to the first monomer molecule to produce the chain initiating species M_1·

$$R \cdot + M \xrightarrow{k_i} M_1 \cdot \quad (2.2)$$

where M represents a monomer molecule and k_i is the rate constant for the initiation step [Eq. (2.2)]. For the polymerization of $CH_2=CHY$, Eq. (2.2) takes the form

$$R \cdot + CH_2=CHY \longrightarrow R-CH_2-\underset{Y}{\overset{H}{C}} \cdot \quad (2.3)$$

The radical R· is often referred to as an initiator radical or a primary radical.

Propagation consists of the growth of M_1· by the successive additions of large numbers (hundreds, and perhaps, thousands) of monomer molecules. Each addition creates a new radical which has the same identity as the one previously, except that it is larger by one monomer unit. The successive additions may be represented in general terms by

$$M_n \cdot + M \xrightarrow{k_p} M_{n+1} \cdot \quad (2.4)$$

where k_p is the rate constant for propagation. Propagation with growth of the chain to high polymer proportions takes place very rapidly. The value of k_p for the most monomers is in the range of $10^2 \sim 10^4$ L/(mol·s). This is a large rate constant——much larger than those usually encountered in chemical reactions.

At some point, the propagating polymer chain stops growing and terminates. Termination with the annihilation of the radical centers occurs by bimolecular reaction between radicals. Two radicals react with each other by *combination (coupling)*

$$\sim\sim CH_2-\underset{Y}{\overset{H}{C}} \cdot + \cdot \underset{Y}{\overset{H}{C}}-CH_2\sim\sim \xrightarrow{k_{tc}} \sim\sim CH_2-\underset{Y}{\overset{H}{C}}-\underset{Y}{\overset{H}{C}}-CH_2\sim\sim \quad (2.5)$$

or, more rarely, by disproportionation in which a hydrogen radical that is beta to one radical center is transferred to another radical center. This results in the formation of two polymer molecules—one saturated and one unsaturated.

$$\sim\sim CH_2-\underset{Y}{\overset{H}{C}} \cdot + \cdot \underset{Y}{\overset{H}{C}}-\underset{H}{\overset{H}{C}}\sim\sim \xrightarrow{k_{td}} \sim\sim CH_2-\underset{Y}{\overset{H}{CH}} + \underset{Y}{\overset{H}{C}}=\overset{H}{C}\sim\sim \quad (2.6)$$

Termination can also occur by a combination of coupling and disproportionation. The two different modes of termination can be represented in general terms by

$$\mathrm{M}_n\cdot + \mathrm{M}_m\cdot \xrightarrow{k_{tc}} \mathrm{M}_{n+m} \quad (2.7)$$

$$\mathrm{M}_n\cdot + \mathrm{M}_m\cdot \xrightarrow{k_{td}} \mathrm{M}_n + \mathrm{M}_m \quad (2.8)$$

where k_{tc} and k_{td} are the rate constants for termination by coupling and disproportionation, respectively. One can also express the termination step by

$$\mathrm{M}_n\cdot + \mathrm{M}_m\cdot \xrightarrow{k_t} \text{dead polymer} \quad (2.9)$$

where the particular mode of termination is not specified and

$$k_t = k_{tc} + k_{td} \quad (2.10)$$

The term *dead polymer* signifies the cessation of growth for the propagation radical. The propagation reaction would proceed indefinitely until all the monomer in a reaction system were exhausted if it were not for the strong tendency toward termination. Typical termination rate constants are in the range of $10^6 \sim 10^8$ L/(mol·s) or orders of magnitude greater than the propagation rate constants. The much greater value of k_t (whether k_{tc} or k_{td}) compared to k_p does not prevent propagation because the radical species are present in very low concentrations and because the polymerization rate is dependent on only the one-half power of k_t.

Equations (2.1) through (2.5) constitute the detailed mechanism of a free radical initiated chain polymerization. The chain nature of the process resides in the propagation step (Eq. (2.4)) in which large numbers of monomer molecules are converted to polymer for each initial radical species produced in the first step (Eq. (2.1)). In order to obtain a kinetic expression for the overall rate of polymerization, it is necessary to assume that k_p and k_t are independent of the size of the radical. This is exactly the same type of assumption which was employed in deriving the kinetics of step polymerization. There is ample experimental evidence which indicates that although radical reactivity depends on molecular size, the effect of size vanishes after the pentamer or hexamer.

Monomer disappears by the initiation reaction (Eq. (2.2) and (2.3)) as well as by the propagation reactions (Eq. (2.4)). The *rate of monomer disappearance*, which is synonymous with the *rate of polymerization*, is given by

$$-\frac{d[\mathrm{M}]}{dt} = R_i + R_p \quad (2.11)$$

where R_i and R_p are the rates of initiation and propagation, respectively. However, the number of monomer molecules reacting in the initiation step is far less than the number in the propagation step for a process producing high polymer. To a very close approximation, the former can be neglected and the polymerization rate is given simply by the rate of propagation

$$-\frac{d[\mathrm{M}]}{dt} = R_p \quad (2.12)$$

The rate of propagation, and therefore, the rate of polymerization, is the sum of many individual propagation steps. Since the rate constants for all the propagation steps are the same, one can express the polymerization rate by

$$R_p = k_p[\mathrm{M}\cdot][\mathrm{M}] \quad (2.13)$$

where $[\mathrm{M}]$ is the monomer concentration and $[\mathrm{M}\cdot]$ is the total concentration of all chain radicals, i.e., all radicals of size $\mathrm{M}_1\cdot$ and larger.

Equation (2.13) for the polymerization rate is not directly usable because it contains a term for the concentration of radicals. Radical concentrations are difficult to measure since they are very low (approximately 10^{-8} molar) and it is therefore desirable to eliminate them from Eq. (2.13). In order to do this, the *steady-state* assumption is made that the concentration of radicals increases initially, but almost instantaneously reaches a constant, steady-state value. The rate of change of the concentration of radicals quickly becomes and remains zero during the course of the polymerization. This is equivalent to stating that the rates of initiation R_i and termination R_t of radicals are equal or

$$R_i = 2k_t[\text{M}\cdot]^2 \qquad (2.14)$$

The theoretical validity of the steady-state assumption has been discussed. Its experimental validity has been shown in many polymerizations. Typical polymerizations achieve a steady-state after an induction period which may be at most a few seconds.

The right side of Eq. (2.14) represents the rate of termination. There is no specification as to whether termination is by coupling or disproportionation since both follow the same kinetic expression. The use of the factor of 2 in the termination rate equation follows the generally accepted convention for reactions destroying radicals in pairs. It is also generally employed for reactions creating radicals in pairs as in Eq. (2.1). In using the polymer literature one should be aware that the factor of 2 is not universally employed. Rearrangement of Eq. (2.14) to

$$[\text{M}\cdot] = \left(\frac{R_i}{2k_t}\right)^{1/2} \qquad (2.15)$$

and substitution into Eq. (2.13) yields

$$R_p = k_p[\text{M}]\left(\frac{R_i}{2k_t}\right)^{1/2} \qquad (2.16)$$

for the rate of polymerization. It is seen that Eq. (2.16) has the significant conclusion of the dependence of the polymerization rate on the square root of the initiation rate. Doubling the rate of initiation does not double the polymerization rate; the polymerization rate is increased only by the factor $\sqrt{2}$. This behavior is a consequence of the bimolecular termination reaction between radicals.

——Odian G. Principles of Polymerization. New York: McGraw-Hill Book Company, 1970. 170

Words and Expressions

kinetics	[kaiˈnetiks]	n.	动力学
propagation	[prɔpəˈgeiʃən]	n.	（链）增长
homolytic dissociation			均裂
successive	[səkˈsesiv]	a.	连续的，接连的，一连串的
annihilation	[ənaiəˈleiʃən]	n.	消失，湮灭
combination	[kɔmbiˈneiʃən]	n.	偶合（终止），结合
disproportionation	[disprɔpɔːʃəˈneiʃən]	n.	歧化（终止）
dead polymer			死聚物，失去活性的聚合物
pentamer	[ˈpentæmə]	n.	五聚体
hexamer	[hekˈsæmə]	n.	六聚体
rate of polymerization			聚合速率
steady-state assumption			稳态假设
validity	[vəˈliditi]	n.	有效（性），正确（性）

UNIT 3 Step-Growth Polymerization

Many different chemical reactions may be used to synthesize polymeric materials by step-growth polymerization. These include esterification, amidation, the formation of urethanes, aromatic substitution, etc. Polymerization proceeds by the reactions between two different functional groups, e. g., hydroxyl and carboxyl groups, or isocyanate and hydroxyl groups.

All step-growth polymerizations fall into two groups depending on the type of monomer (s) employed. The first involves two different polyfunctional monomers in which each monomer possesses only one type of functional group. A *polyfunctional monomer* is one with two or more functional groups per molecule. The second involves a single monomer containing both types of functional groups. The synthesis of polyamides illustrates both groups of polymerization reactions. Thus, polyamides can be obtained from the reaction of diamines with diacids

$$n H_2N-R-NH_2 + n \ HO_2C-R'-CO_2H \longrightarrow$$
$$H-(NH-R-NHCO-R'-CO)_n-OH + (2n-1)H_2O \quad (3.1)$$

or from the reaction of amino acids with themselves

$$n H_2N-R-CO_2H \longrightarrow H-(NH-R-CO)_n-OH + (n-1)H_2O \quad (3.2)$$

The two groups of reactions can be represented in a general manner by the equations as follows

$$A-A + B-B \longrightarrow -[A-A-B-B]-$$
$$A-B \longrightarrow -[A-B]-$$

Reaction (3.1) illustrates the former, while (3.2) is of the latter type.①

Polyesterification, whether between diol and dibasic acid or intermolecularly between hydroxy acid molecules, is an example of a step-growth polymerization process. The esterification reaction occurs anywhere in the monomer matrix where two monomer molecules collide, and once the ester has formed, it, too, can react further by virtue of its still-reactive hydroxyl or carboxyl groups.② The net effect of this is that monomer molecules are consumed rapidly without any large increase in molecular weight.③ Fig. 3.1 illustrates this phenomenon. Assume, for example, that each square in Fig. 3.1 represents a molecule of hydroxy acid. After the initial dimer molecules form (b), half the monomer molecules have been consumed and the average degree of polymerization ($\overline{DP}$) of polymeric species is 2. As trimer and more dimer molecules form (c), more than 80% of the monomer molecules have reacted, but $\overline{DP}$ is still only 2.5. When all the monomer molecules have reacted (d), $\overline{DP}$ is 4. But each polymer molecule that forms still has reactive end groups; hence the polymerization reaction will continue in a stepwise fashion, with each esterification step being identical in rate and mechanism to the initial esterification of monomers.④ Thus, molecular weight increases slowly even at high levels of monomer conversion, and it will continue to increase until the viscosity build-up makes it mechanically too difficult to remove water of esterification or for reactive end groups to find each other.

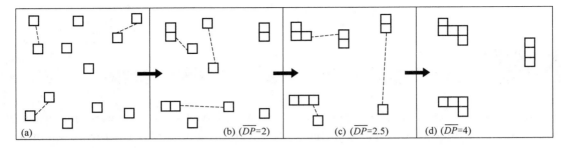

Fig. 3.1

Difference in behavior on heating of a low molecular weight compound (benzene) and a polymer (polyethylene)

It can also be shown that in the A—A + B—B type of polymerization, an exact stoichiometric balance is necessary to achieve high molecular weights. If some monofunctional impurity is present, its reaction will limit the molecular weight by rendering a chain end inactive. Similarly, high-purity monomers are necessary in the A—B type of polycondensation, and it follows that high-yield reactions are the only practical ones for polymer formation, since side reactions will upset the stoichiometric balance.

——Stevens M P. Polymer Chemistry. London: Addison-Wesley Publishing Company, 1975. 13

Words and Expressions

step-growth polymerization			逐步聚合
synthesize	['sinθisaiz]	v.	合成
esterification	[estərifi'keiʃən]	n.	酯化（作用）
amidation	[əmi'deiʃən]	n.	酰胺化（作用）
urethane	['juəriθein]	n.	氨基甲酸酯
aromatic	[ærəu'mætik]	a.	芳香（族）的
substitution	[sʌbsti'tjuːʃən]	n.	取代，代替
functional group			官能团
hydroxyl	[hai'drɔksil]	n.	羟基
carboxyl	[kɑː'bɔksil]	n.	羧基
isocyanate	[aisəu'saiəneit]	n.	异氰酸酯
polyfunctional	[pɔli'fʌŋkʃənl]	a.	多官能度的
synthesis	['sinθisis]	n.	合成
polyamide	[pɔli'æmaid]	n.	聚酰胺
diamine	['daiəmin]	n.	二（元）胺
diacid	[dai'æsid]	n.	二（元）酸
amino	[ə'minəu]	n.	氨基
		a.	氨基的
polyesterification	[pɔliestərifi'keiʃən]	n.	聚酯化（作用）
diol	['daiɔːl]	n.	二（元）醇
dibasic	[dai'beisik]	a.	二元的
hydroxy acid			羟基酸
dimer	['daimə]	n.	二聚物［体］
trimer	['traimə]	n.	三聚物［体］
reactive	[ri'æktiv]	a.	反应性的，活性的
molecular weight			分子量
conversion	[kən'vəːʃən]	n.	转化率
stoichiometric	[stɔikiə'metrik]	a.	当量的，化学计算量的
monofunctional	[mɔnə'fʌŋkʃənl]	a.	单官能度的
impurity	[im'pjuːəriti]	n.	杂质
purity	['pjuːəriti]	n.	纯度
polycondensation	[pɔlikɔnden'seiʃən]	n.	缩（合）聚（合）
yield	[jiːld]	n.	产率
side reaction			副反应

Phrases

fall into 归入，可分为
depending on 根据
in a general manner 一般来说［讲］
be of 具有

by virtue of 依靠，借助于
in a stepwise fashion 以逐步的方式，逐步地
be identical in...to... 在…方面和…是相同的
build-up 增大

Notes

① "Reaction (3.1) illustrates the former, while (3.2) is of the latter type." 逐步聚合反应有前后两种形式，"the former" 代表前一种形式，而 "the latter type" 为后一种形式。句子后一部分为由 "while" 引导的并列分句，在该从句中 "(3.2)" 前省略了 "Reaction"。"is of" 意为 "具有"。全句译文为："反应（3.1）说明前一种形式，而反应（3.2）具有后一种形式。"

② "The esterification reaction occurs anywhere in the monomer matrix where two monomer molecules collide, and once the ester has formed, it, too, can react further by virtue of its still-reactive hydroxyl or carboxyl groups." 该句子为并列复合句，用 "and" 把两个分句分开。在前一分句中的 "anywhere" 为一个不定代词，意为 "任何（一个）地方"，与 "anyplace" 为同义词，其后由 "where" 引导的句子为 "anywhere" 的定语从句，可译为 "在单体本体中两个单体分子相碰撞的任何位置"。在后一分句中，主句为 "it, too, can react further"，主句前面为 "once" 引导的时间状语从句，主句后面的 "by virtue of…" 短语为主句的方式状语。全句译文："酯化反应出现在单体本体中两个单体分子相碰撞的任何位置，且酯一旦形成，依靠酯上仍有活性的羟基或羧基还可以进一步进行反应。"

③ "The net effect of this is that monomer molecules are consumed rapidly without any large increase in molecular weight." "This" 是指以上所说的单体或聚合物末端基团间均可进行酯化作用这一事实，"the net effect" 意为最终结果，即后果，为主语。"is" 之后为表语从句，从句中 "without" 引导的介词短语为从句的状语。全句译文："（单体或聚合物末端活性基团间）酯化的结果是单体分子很快地被消耗掉，而（聚合物的）分子量却没有增大多少。"

④ "But each polymer molecule that forms still has reactive end groups; hence the polymerization reaction will continue in a stepwise fashion, with each esterification step being identical in rate and mechanism to the initial esterification of monomers." "hence" 前后为两个并列的句子，在前一个句子中 "that forms" 为主语 "each polymer molecule" 的定语从句；在后一个句子中，"with" 引导的介词短语为方式状语，介词 "with" 后为一分词独立结构。全句译文："但所形成的每个聚合物分子仍然具有反应性末端基团，因此，聚合反应将以逐步的方式继续进行，其每一步酯化反应的反应速率和反应机理均与初始单体的酯化作用相同。"

Exercises

1. *Translate the following passage into Chinese*

The polymerization rate may be experimentally followed by measuring the changes in any of several properties of the system such as density, refractive index, viscosity, or light absorption. Density measurements are among the most accurate and sensitive of the techniques. The density increases by 20~25 percents on polymerization for many monomers. In actual practice the volume of the polymerizing system is measured by carrying out the reaction in a *dilatometer*. This is specially constructed vessel with a capillary tube which allows a highly accurate measurement of small volume changes. It is not uncommon to be able to detect a few hundreds of a percent polymerization by the dilatometer technique.

2. *Write out an abstract in English for the text in this Unit*

3. *Put the following words into English*

聚氯乙烯　烯烃　溶解　溶胀　机理　自由基　引发剂　苯乙烯　吸收　氯　活化能

4. *Put the following words into Chinese*
   chain polymerization　　phosgene　　titanium trichloride　　metal complex　　propagation
   disproportionation　　polyvinyl alcohol　　common salt　　fragment　　linear polymer
   branched polymer　　copolymer

Reading Materials

Polyethers

1. Acetals

Polyoxymethylene, —O—CH$_2$—O—CH$_2$—, which is prepared from formaldehyde or trioxane, is a polymer of extraordinary importance. Although Staudinger studied these thermally stable polymers in detail some 50 years ago, they have not gained commercial significance until more recently.

Major industrial applications include use in place of nonferrous metals and, more general, in the field of communication equipment, automotive, plumbing, and hardware. The outstanding properties of acetal homopolymers include high strength and rigidity, excellent dimensional stability and resilience, and low hysteresis losses. Together with nylon, acetals have the highest abrasion resistance of all commercial thermoplastics. Their low coefficient of friction (0.1 to 0.3) makes them desirable materials for bearing and other moving parts.

Acetal copolymers based on trioxane excel in their resistance to creep under load at elevated temperatures and are, in this respect, comparable to die cast metal. The densities of the homopolymer and copolymer are 1.425 and 1.42 g/cm^3, respectively.

2. Poly (oxyethylene) glycols [poly (ethylene oxide)]

Poly (oxyethylene) glycols or poly (ethylene oxide) are hydrophilic in nature and range in appearance from colorless liquids to wax-like solids, depending on molecular weight. They have moderate strength and stiffness; their crystalline melting point is ca. 65℃. These polymers are useful in applications where water-soluble films, fibers, etc. are required, and have also found wide use as thickeners, additives in glues and sizing systems for fibers, and in pharmaceutical packaging.

3. Poly (oxypropylene) glycols [Poly (propylene ethers)]

The commercial species are generally prepared by reaction of propylene oxide with propylene glycol. They are widely used in the preparation of urethane polymers. While these polypropylene ethers are endowed with pendant methyl groups,

$$HO-CH-CH_2-O \left[CH_2-CH-O \right]_n CH_2-CH-OH$$
$$\quad\ \ |\qquad\qquad\qquad\ \ |\qquad\qquad\quad\ \ |$$
$$\quad\ CH_3\qquad\qquad\qquad CH_3\qquad\qquad\ \ CH_3$$

those prepared from oxetanes (trimethylene oxides) are essentially linear in nature.

4. Poly (oxytetramethylene) glycols [Polybutylene ethers]

These compounds are prepared from polymerization of tetrahydrofuran; they are also linear, while those prepared from butylene oxide are marked by pendant groups, depending on the ratio of 1,2- and 2,3-butylene oxide employed.

5. Poly (oxypropylene) triols

These triols generally manufactured by reaction of propylene oxide with low molecular weight triols, such as trimethylolpropane, glycerol, or 1,2,6-hexanetriol, are among the most important class of polyethers used in the manufacture of urethane polymers.

6. Nitrogen-Containing Polyethers

These polyethers used in the manufacture of some types of urethane polymers, can be prepared by reacting cyclic oxides with aqueous ethylene-diamine using alkali hydroxide as a catalyst. Other nitrogen containing initiators are used, occasionally, e. g., benzenesulfonamide, 2-aminoethylethanolamine, *N*-methyldiethanolamine, diethylenetriamine, and tris (hydroxymethyl) aminomethane.

7. Miscellaneous Polyethers

Copolymers of styrene oxide and various monomers such as epichlorohydrin, ethylene oxide, propylene oxide, cyclohexene oxide, and butadiene monoxide have been prepared and are commercially available. Alkylene carbonates have also been used to bring about oxyalkylation. Polymers from 1,4-epoxycyclohexane with tetrahydrofuran, and other cyclic oxides have also been prepared.

8. Chlorinated Polyethers

Chlorinated polyethers, or polydichloromethyl oxacyclobutane are characterized by high thermal and chemical stability resulting from the molecular structure

$$\left[-CH_2-\underset{\underset{CH_2Cl}{|}}{\overset{\overset{CH_2Cl}{|}}{C}}-CH_2-O- \right]_n$$

in which chlorine represents about 46% of the polymer's weight. These polymers are linear and crystalline, have good impact strength and low moisture absorption. They are chiefly employed as liners for pipes and tubing, and in the manufacture of monofilaments.

9. Polythioethers

Polythioethers usually prepared by the acid-catalyzed condensation of thiodiglycol with itself or with polyols, have not attained commercial significance; this is partly due to the difficulty in eliminating all objectionable odors.

——Boening H V. Structure and Properties of Polymers. London: Geog Publishers Stuttgart, 1973. 30

Words and Expressions

polyether	[ˌpɔliˈiːθə]	n.	聚醚
acetal	[ˈæsitæl]	n.	缩醛
polyoxymethylene	[ˌpɔliɔksiˈmeθiliːn]	n.	聚甲醛
trioxane	[traiˈɔksein]	n.	三聚甲醛，三噁烷，（均）三氧环己烷
nonferrous metal			有色金属
hysteresis loss			滞后损耗［失］
coefficient of friction			摩擦系数
resistance to creep			抗蠕变性
polyethylene oxide			聚氧化乙烯
hydrophilic	[ˌhaiˈdrɔfilik]	a.	亲水的
thickener	[ˈθikənə]	n.	增稠剂
glue	[gluː]	n.	胶水，胶黏剂
sizing system			上浆剂，涂饰剂
oxetane	[ˈɔksətein]	n.	氧杂环丁烷
triol	[ˈtraiɔl]	n.	三元醇
1,2,6-hexanetriol	[hekˈseinˈtraiɔl]	n.	1,2,6-三羟基己烷
ethylenediamine	[ˈeθiliːnˈdaiəmiːn]	n.	乙二胺
alkali hydroxide			强碱性的氢氧化物
benzenesulfonamide	[ˈbenzinˈsʌlfəunæmaid]	n.	苯磺酰胺

2-aminoethylethanolamine	[ə'minəu'eθənɔləmi:n]	n.	2-氨乙基乙醇胺
N-methyldiethanolamine	['meθildai'eθənɔləmin]	n.	N-甲基二乙醇胺
diethylenetriamine	[dai'eθili:n'traiəmi:n]	n.	二亚乙基三胺
tris (hydroxymethyl) aminomethane	[tris haidrɔksimeθil ə'min əu'eθein]	n.	三羟甲基氨基甲烷
miscellaneous	[misi'leiniəs]	a.	各种各样的，其他的
epichlorohydrin	[epiklɔ:rə'haidrin]	n.	环氧氯丙烷，表氯醇
butadiene monoxide		n.	丁二烯单氧化物
oxyalkylation	[ɔksiælki'leiʃən]	n.	烷氧基化（作用）
chlorinated polyether		n.	氯化聚醚
polydichloromethyl oxacyclobutane	[pɔlidai'klɔ:rəmeθil ɔksəsaikləu'bju:tein]	n.	聚二氯甲基氧杂环丁烷
liner	['lainə]	n.	衬里
monofilament	[mɔnəu'filəmənt]	n.	单丝
polyol	['pɔliɔl]	n.	多元醇
odor	['əudə]	n.	气味

UNIT 4　Ionic Polymerization

Ionic polymerization, similar to radical polymerization, also has the mechanism of a chain reaction. The kinetics of ionic polymerization are, however, considerably different from that of radical polymerization.

(1) The initiation reaction of ionic polymerization needs only a small activation energy. Therefore, the rate of polymerization depends only slightly on the temperature. Ionic polymerizations occur in many cases with explosive violence even at temperatures below 50℃ (for example, the anionic polymerization of styrene at −70℃ in tetrahydrofuran, or the cationic polymerization of isobutylene at −100℃ in liquid ethylene).

(2) With ionic polymerization there is no compulsory chain termination through recombination, because the growing chains can not react with each other. ① Chain termination takes place only through impurities, or through the addition of certain compounds such as water, alcohols, acids, amines, or oxygen, and in general through compounds which can react with polymerizing ions under the formation of neutral compounds or inactive ionic species. ② If the initiators are only partly dissociated, the initiation reaction is an equilibrium reaction, where reaction in one direction gives rise to chain initiation and in the other direction to chain termination. ③

In general ionic polymerization can be initiated through acidic or basic compounds. For cationic polymerization, complexes of BF_3, $AlCl_3$, $TiCl_4$, and $SnCl_4$ with water, or alcohols, or tertiary oxonium salts have shown themselves to be particularly active. The positive ions are the ones that cause chain initiation. For example:

$$BF_3 + HO-R \rightleftharpoons \begin{bmatrix} F \\ | \\ F-B-F \\ | \\ OR \end{bmatrix}^{(-)} + H^{(+)} \rightleftharpoons H[BF_3OR]$$

$$[(C_2H_5)_3O]\,BF_4 \rightleftharpoons (C_2H_5)_3O^{(+)} + BF_4^{(-)}$$
triethyloxonium-borofluoride

However, also with HCl, H_2SO_4, and $KHSO_4$, one can initiate cationic polymerization. Initiators for anionic polymerization are alkali metals and their organic compounds, such as phenyllithium, butyllithium, phenyl sodium, and triphenylmethyl potassium, which are more or less strongly dissociated in different solvents. To this group belong also the so called Alfin catalysts, which are a mixture of sodium isopropylate, allyl sodium, and sodium chloride. ④

With BF_3 (and isobutylene as the monomer), it was demonstrated that the polymerization is possible only in the presence of traces of water or alcohol. If one eliminates the trace of water, BF_3 alone does not give rise to polymerization. Water or alcohols are necessary in order to allow the formation of the BF_3-complex and the initiator cation according to the above reactions. However, one should not describe the water or the alcohol as a "cocatalyst".

Just as by radical polymerization, one can also prepare copolymers by ionic polymerization, for example, anionic copolymers of styrene and butadiene, or cationic copolymers of isobutylene and styrene, or isobutylene and vinyl ethers, etc. As has been described in detail with radical polymerization, one can characterize each monomer pair by so-called reactivity ratios r_1 and r_2. ⑤ The actual values of these two parameters are, however, different from those used for radical copolymerization.

——Vollmert B. Polymer Chemistry. Berlin: Springer-Verlag, 1973. 163

Words and Expressions

ionic polymerization			离子型聚合
radical polymerization			自由基型聚合
kinetics	[kai'netiks]	n.	动力学
anionic	[ænai'ɔnik]	a.	阴［负］离子的
cationic	[kætai'ɔnik]	a.	阳［正］离子的
tetrahydrofuran	[tetrəhaidrɔ'fju:ræn]	n.	四氢呋喃
isobutylene	[aisəu'bju:tili:n]	n.	异丁烯
chain termination			链终止
growing chain			生长链，活性链
ion	['aiən]	n.	离子
neutral	['nju:trəl]	a.	中性的
dissociate	[di'səuʃieit]	v.	离解
equilibrium	[i:kwi'libriəm]	n.	平衡
tertiary	['tə:ʃəri]	a.	三元的，叔［特］的
oxonium	[ɔk'səunjəm]	n.	氧鎓
positive	['pɔzitiv]	a.	正的，阳（性）的
triethyloxonium-borofluoride			三乙基氧鎓硼氟酸盐
alkali metal			碱金属
phenyllithium	[fenil'liθiəm]	n.	苯基锂
butyllithium	[bju:til'lθiəm]	n.	丁基锂
phenyl sodium			苯基钠
triphenylmethyl potassium			三苯甲基钾
Alfin catalyst			醇（碱金属）烯催化剂
isopropylate	[aisəu'prəupilei]	n.	异丙醇金属，异丙氧化金属
allyl	['ɔ:lil]	n.	烯丙基
cation	['kætaiən]	n.	正［阳］离子
cocatalyst	[kəu'kætəlist]	n.	助催化剂
copolymer	[kəu'pɔlimə]	n.	共聚物
vinyl ether			乙烯基醚
characterize	['kærikəraiz]	v.	表征，成为…的特征
reactivity ratio			竞聚率
parameter	[pə'ræmitə]	n.	参数
copolymerization	[kəupɔliməri'zeiʃən]	n.	共聚［合］

Phrases

with explosive violence　极其猛烈地
give rise to…　引起，导致，产生，使…发生
more or less　近乎，大体上，在不同程度上
so（-）called　所谓的

in the presence of…　在…存在下
just as　正与……一样，正当…时候
in detail　详细地

Notes

① "With ionic polymerization there is no compulsory chain termination through recombination, because the growing chains can not react with each other." 在此句中"with"作"对于"解。"compulsory chain termination"译为"强迫链终止"。"through"引导的介词短语为方式状语。"because"后为原因状语从

句。全句译文："对于离子型聚合来说，不存在通过再结合反应而进行的强迫链终止，因为生长链之间不能发生链终止反应。"

② "Chain termination takes place only through impurities, or through the addition of certain compounds such as water, alcohols, acids, amines, or oxygen, and in general through compounds which can react with polymerizing ions under the formation of neutral compounds or inactive ionic species." 主句的主语为 "chain termination"，谓语是 "takes place"，其后有三个由 "through" 引导的介词短语为方式状语。"which" 引导的从句是 "compounds" 的定语从句，在这个从句中的 "under the formation of…" 意为"在…期间"或"在…过程中"。句末 "species" 指"聚合物"。全句译文："链终止反应仅仅通过杂质而发生，或者说通过和某些像水、醇、酸、胺或氧这样的化合物进行加成反应而发生，且一般来说（链终止反应）可通过这样的化合物来进行，这种化合物在中性聚合物或没有聚合活性的离子型聚合物生成的过程中可以和活性聚合物离子进行反应。"

③ "If the initiators are only partly dissociated, the initiation reaction is an equilibrium reaction, where reaction in one direction gives rise to chain initiation and in the other direction to chain termination." 在 "if" 引导的条件状语从句之后为主句，在主句的表语之后为其由 "where" 引导的非限定性定语从句。在 "in the other direction" 之后，为避免重复而省略了 "gives rise"。全句译文："如果引发剂仅为部分地离解，引发反应即为一个平衡反应，在出现平衡反应的场合，在一个方向上进行链引发反应，而在另一个方向上则发生链终止反应。"

④ "To this group belong also the so called Alfin catalysts, which are a mixture of sodium isopropylate, allyl sodium, and sodium chloride." 主句是一个倒装句，为了强调宾语 "to this group"，故将其提前。"which" 后为 "Alfin catalysts" 的非限定性定语从句。全句译文："所谓的 Alfin 催化剂就是属于这一类，这类催化剂是异丙醇钠、烯丙基钠和氯化钠的混合物。"

⑤ "As has been described in detail with radical polymerization, one can characterize each monomer pair by so-called reactivity ratios r_1 and r_2." 在该句中 "one" 是主句的主语，作"人们"解。主句前为由 "as" 引出的非限定性定语从句，用来说明整个主句，"as" 在主句中作主语。全句译文："正如对自由基型聚合已经详细描述过的那样，人们可以用所谓的竞聚率 r_1 和 r_2 来表征每个单体对。"

Exercises

1. *Make a sentence for each phrase as follows*
   (1) to give rise to…
   (2) to be made up of…
   (3) to be attributed to…
   (4) to be characteristic of…

2. *Translate the following passage into English*
   合成聚合物在各个领域中起着与日俱增的重要作用，聚合物通常是由单体通过加成聚合与缩合聚合制成的。就世界上的消耗量而论，聚烯烃和乙烯基聚合物居领先地位，聚乙烯、聚丙烯等属聚烯烃，而聚氯乙烯、聚苯乙烯等则为乙烯基聚合物。聚合物可广泛地用做塑料、橡胶、纤维、涂料、胶黏剂等。

3. *Put the following words into Chinese*
   diamine   polyfunctional monomer   hydroxyl group   ultrasonics   acetal   nitrogen-containing   polyether   crystal   irradiation   saturated salt solution   kinetic chain length   homolytic dissociation

4. *Put the following words into English*
   合成 氢 酯化 羧基 分解 双键 现象 浓度 催化剂 聚苯乙烯 逐步聚合 二聚体

Reading Materials

Review of Ionic Polymerization

Ionic polymerization received prominence about 35 years ago when isobutylene was commercially polymerized by two processes which, with some modification, are still used today. One process uses aluminum chloride as the initiator and the other uses boron trifluoride; both cationic polymerization processes are carried out at low temperatures. A number of additional commercial processes based on cationic and anionic polymerizations have been developed. Cyclic ethers, e. g. tetrahydrofuran, are polymerized cationically to relatively low molecular weight hydroxyl terminated polyethers which have found important uses in polyurethanes. Trioxane is copolymerized with a small amount of ethylene oxide to form a useful copolymer of polyoxymethylene. Other products which are of interest are the polymers of caprolactone and epichlorohydrin and polymers of various epoxides, mainlythose of glycidyl ethers which are most commonly known as epoxy resins. Anionic polymerization on a commercial scale has developed along the lines of styrene and isoprene polymers. Stereorubber, stereoregular 1,4-*cis* isoprene, are based on lithium initiators and were introduced in the middle 1950s. Triblock polymers based on A—B—A block polymers of isobutylene with styrene as endblocks and prepared from living polymers have been known since the early 1960s.

Some of these developments in ionic polymerization were the results of fundamental scientific investigations, but others, particularly the earlier work, were based on intuitive work of ingenious inventors.

It is clear that the general scope of ionic polymerizations involves a number of different fields of interest. It is the purpose of this symposium to bring together investigators involved in investigations of various disciplines in order to disseminate the information and cross-fertilize the ideas developed in their own area of interest.

Ionic polymerization can be looked upon from the traditional mechanistic or historical as well as from the practical point of view, and it has been treated in numerous ways. In cationic polymerization, isobutylene polymerization has often been compared with styrene polymerization; carbenium ion (carbocation) polymerization has competed or coexisted with oxonium polymerization; and coordinative ionic polymerization has profited from classical ionic polymerization.

Cationic and anionic polymerization are chain growth polymerization which frequently involve bond-opening reactions of carbon-carbon double bonds with carbenium (carbocations) or carbanions as polymeric propagating species. Heteroatom-containing monomers, such as aldehydes, polymerize by bond-opening polymerization of carbonyl groups, and cyclic ethers by ring-opening polymerizations with oxonium ions or alkoxide ions at the end of the growing polymer chain. Lactam and lactone polymerizations involve the ring opening of the carbonyl group, but the methylene group attached to the ether oxygen may also be the point of attack.

In addition to the above-mentioned chain growth polymerization to form high molecular weight polymers by ionic processes, polymer formation by a step growth polymer process which involves ionic intermediates has frequently been investigated. It involves electrophilic and nucleophilic substitution reactions on aliphatic or aromatic compounds to form high molecular weight polymers. Discussions of these polymerizations are usually not included when ionic polymerization is discussed.

Cationic polymerization of olefins is carried out with olefins which are substituted with electron-donating groups. The best studied examples are isobutylene, styrene, butadiene, and isoprene. Methyl substitution, especially disubstitution on one carbon atom, as for example in isobutylene, causes the carbon-carbon double bonds to be properly polarized for the addition of the electrophile.

Cationic carbenium ion polymerization is the most easily discussed with isobutylene as the example. The electrophile, for example a proton, acts as the initiator, and termination of the growing polymeric cation is primarily by proton transfer to another isobutylene molecule to form an olefin-terminated polyisobutylene and a new tertiary butyl carbenium ion which is capable of the further propagation.

Initiation: $R^{(+)} + CH_2=C(CH_3)_2 \longrightarrow R-CH_2-\overset{(+)}{C}(CH_3)_2$

Transfer:

$$P-CH_2-\overset{(+)}{C}(CH_2)_3 \longrightarrow P-CH=C(CH_3) + H^{(+)}$$

$$+ CH_2=C(CH_3)_2 \quad \text{or} \quad P-CH_2-\underset{CH_3}{\overset{|}{C}}=CH_2 + H^{(+)}$$

$$P-CH=C(CH_3)_2 + (CH_3)_3C^{(+)}$$

Fundamental knowledge has been acquired about cationic carbenium ion polymerization over the years, particularly the influence of coinitiators, the influence of temperatures on molecular weight, and, most recently, the importance of the equilibrium between carbenium ions and nonionic species as in the case of the polymerization of styrene with perchloric acid. A number of other examples of equilibrium between ionic and nonionic intermediates have recently become known in the ring-opening polymerization of heterocyclic monomers.

A parallel to the proton initiation and proton transfer in cationic olefin polymerization is the fluoride ion initiation and fluoride ion transfer in anionic fluoroolefin polymerization, for example, in the case of the oligomerization of hexafluoropropylene. The initiating fluoride ion adds to the less hindered CF_2 group of the olefin to form the heptafluoroisopropyl anion which is then capable of adding a new molecule of hexafluoropropylene. Because of the relative ease of fluoride loss under these conditions, polymerization to high molecular weight does not occur and the formation of highly branched dimeric and trimeric fluoroolefins is the result.

$$F^{(-)} + CF_2=CF-CF_3 \longrightarrow CF_3\overset{(-)}{C}FCF_3$$

$$(CF_3)_2CF^{(-)} + CF_2=CF-CF_3 \longrightarrow (CF_3)_2CF-CF_2\overset{(-)}{C}F-CF_3$$

$$\downarrow -F^{(-)}$$

$$(CF_3)_2CF-CF=CF-CF_3$$

The reaction is very similar to the oligomerization of propylene in the presence of protic acid.

——Furukawa J. Vogl O. Ionic Polymerization, New York: Marcel Dekker, 1976. 1

Words and Expressions

modification	[mɔdifiˈkeiʃən]	n.	改性
boron trifluoride			三氟化硼
caprolactone	[kæprəuˈlæktɔn]	n.	己内酯，羟乙基己酸内酯
epoxide	[eˈpɔksaid]	n.	环氧化合物
glycidyl ether			缩水甘油醚
isoprene	[ˈaisəupriːn]	n.	异戊二烯
stereoregular	[stiəriəuˈregjulə]	a.	立构规整性的，有规立构的
block polymer			嵌段聚合物
endblock	[endˈblɔk]	n.	封端（聚合物）
symposium	[simˈpəuziəm]	n.	（专题）讨论会，论文集
carbenium ion			碳正离子，碳鎓离子
coordinative polymerization			配位（离子）聚合
carbocation	[ˈkɑːbəkætaiəu]	n.	碳正离子
carbanion	[ˈkɑːbənaiən]	n.	碳负离子
carbonyl group			羰基
alkyloxide ion			烷氧基离子
lactam	[ˈlæktæm]	n.	内酰胺
lactone	[ˈlæktəun]	n.	内酯
intermediate	[intəˈmiːdjət]	n.	中间产物，中间体
electrophilic substitution			亲电取代
nucleophilic substitution			亲核取代
aliphatic compound			脂肪族化合物
aromatic compound			芳香族化合物
proton	[ˈprəutɔn]	n.	质子
nonionic	[nɔnaiˈɔnik]	a.	非离子的
oligomerization	[ɔligəuməraizeiʃən]	n.	低[齐]聚物（作用）
hexafluoropropylene	[heksəˈfluərəˈprəupiliːn]	n.	六氟丙烯
protic acid			质子酸

UNIT 5 Introduction to Living Radical Polymerization

Traditional methods of living polymerization are based on ionic, coordination or group transfer mechanisms. Ideally, the mechanism of living polymerization involves only initiation and propagation steps. All chains are initiated at the commencement of polymerization and propagation continues until all monomer is consumed.

A type of novel techniques for living polymerization, known as living (possibly use "controlled" or "mediated") radical polymerization, is developed recently. The first demonstration of living radical polymerization and the current definition of the process can be attributed to Szwarc. Up to now, several living radical polymerization processes, including atom transfer radical polymerization (ATRP), reversible addition-fragmentation chain transfer polymerization (RAFT), nitroxide-mediated polymerization (NMP), etc., have been reported one after another.

The mechanism of living radical polymerization is quite different not only from that of common radical polymerization but also from that of traditional living polymerization. It relies on the introduction of a reagent (active species) that undergoes reversible termination with the propagating radicals thereby converting them to a following dormant form:[①]

$$\text{propagating radical} \qquad \text{active species} \qquad \text{dormant species}$$

$$\left[\text{I}-\text{CH}_2-\underset{Y}{\overset{X}{C}}- \right]_n -\text{CH}_2-\underset{Y}{\overset{X}{C}}\cdot + \cdot Z \rightleftharpoons \left[\text{I}-\text{CH}_2-\underset{Y}{\overset{X}{C}}- \right]_n -\text{CH}_2-\underset{Y}{\overset{X}{C}}-Z$$

The specificity in the reversible initiation-termination step is of critical importance in achieving living characteristics.[②] This enables the active species concentration to be controlled and thus allows such a condition to be chosen that all chains are able to grow at a similar rate (if not simultaneously) throughout the polymerization.[③] This has, in turn, enabled the synthesis of polymers with controlled composition, architecture and molecular weight distribution. They also provide routes to narrow dispersity end-functional polymers, to high purity block copolymers, and to stars and other more complex architecture.

The first step towards living radical polymerization was taken by Ostu and his colleagues in 1982. In 1985, this was taken one step further with the development by Solomon *et al.* of nitroxide-mediated polymerization (NMP). This work was first reported in the patent literature and in conference papers but was not widely recognized until 1993 when Georges *et al.* applied the method in the synthesis of narrow polydispersity polystyrene. The scope of NMP has been greatly expended and new, more versatile, methods have appeared. The most notable methods are atom transfer radical polymerization (ATRP) and polymerization with reversible addition fragmentation (RAFT). Up to 2000, this area already accounted for one third of all papers in the field of radical polymerizaion, as shown in Fig. 5.1. Naturally, the rapid growth of the number of the papers in the field since 1995 ought to be almost totally attributable to development in this area.[④]

Fig. 5.1

Publication rate of journal papers on radical polymerization and on living radical polymerization for period 1975-2002

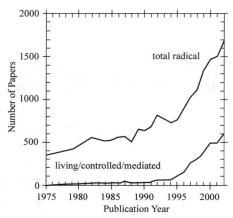

——Moad G,Solomon D H. The Chemistry of radical polymerization,Second Fully Revised Edition,New York:Elsevier Ltd,2006

Words and Expressions

ideally	[ai'diəli]	adv.	理想地；理论上
group transfer mechanism			基团转移机理
commencement	[kə'mensmənt]	n.	开始
living polymerization			活性聚合
living radical polymerization			活性自由基聚合
simultaneous reaction			平行反应
undergo	[ˌʌndə'gəu]	v.	进行，经历
mediate	['mi:dieit]	v.	传递，媒介
specificity	[ˌspesi'fisəti]	n.	特异性，专一性，特征
critical	['kritikəl]	a.	决定性的
nitroxide-mediated polymerization（NMP）			硝基氧介导聚合
atom transfer radical polymerization（ATRP）			原子转移自由基聚合
reversible addition-fragmentation chain transfer polymerization（RAFT）			可逆加成-断裂链转移聚合
architecture	['ɑ:kitektʃə]	n.	结构
dormant	['dɔ:mənt]	a.	休眠的
dormant species			休眠种
active species			活性种
versatile	['və:sətail]	a.	用途广的
simultaneously	[ˌsaiməl'teiniəsli]	adv.	平行地，联立地

Phrases

be attributed to　归功于,归因于
be of + 抽象名词　具有…（属性）
rely on　依赖,依靠
convert... to...　把…转化成…
enable... to do...　能使…做…

allow... to do...　使…做…
take the first step　迈出第一步,开头（做某事）
account for　占,证明
be attributable to　归功于,归因于

Notes

① "It relies on the introduction of a regent that undergoes reversible termination with the propagating radicals thereby converting them to a following dormant form." 该句中 "It" 是主语,指的是 "living radical polymerization"。"relies on" 是谓语,即 "依赖于"。"that" 引导的从句作名词 "reagent" 的定语。最后的 "converting…" 分词短语作主句的伴随状语。本句译文: "活性自由基聚合依赖于向体系中引入一种可以和增长自由基进行可逆终止的试剂,形成休眠种。"

② "The specificity in the reversible initiation-termination step is of critical importance in achieving living characteristics."该句的主语是"The specificity in the reversible initiation-termination",从字面上应翻译成"可逆-引发终止反应的特性",但是若翻译成"这种特殊的可逆引发-终止反应"似乎更能反映其内涵。"is of + 抽象名词"作"具有（某种属性）"解。本句译文："这种特殊的可逆引发-终止反应对于获得分子链活性来说具有决定性的重要意义。"

③ "This enables the active species concentration to be controlled and thus allows such a conditions to be chosen that all chains are able to grow at a similar rate (if not simultaneously) throughout the polymrization."该句中"This"是主语，指的是"the reversible initiation-termination",即"可逆引发终止反应"。其后用了"enable... to do..."和"allow... to do..."结构，意指"使…（能够）做什么"。最后是"such... that..."引导的结果从句。本句译文："可逆引发终止使活性中心的浓度能够得以控制，这样就可以来选择适宜的反应条件，使得在整个聚合反应过程中（只要没有平行反应）所有的分子链都能够以相同的速度增长。"

④ Naturally, the rapid growth of the number of the papers in the field since 1995 ought to be almost totally attributable to development in this area. 本句译文："很自然，自从1995年以来，在这个领域里论文数量的快速增长应当完全归功于这个领域的发展。"

Exercises

1. *Traslate the following passage into Chinese*

   The first step towards living radical polymerization was taken by Ostu and his colleagues in 1982. In 1985, this was taken one step further with the devel opment by Solomon *et al*. of Nitroxide-Mediated Polymerization (NMP). This work was first reported in the patent literature and in conference papers but was not widely recognized until 1993 when Georges *et al*. applied the method in the synthesis of narrow polydispersity polystyrene.

2. *Put the following words into Chinese*

   group transfer mechanism   commencement
   controlled radical polymerization   definition
   traditional living polymerization   patent literature
   reversible initiation-termination step account for
   dormant species   specificity
   novel technique   architecture
   be attributed to   rely on
   controlled composition   demonstration

3. *Put the following words into Chinese*

   可逆反应 聚苯乙烯 理想地 自然地 转化
   多分散性 活性中心 平行的 定义   试剂

Reading Materials

How Do Scientists Design New Polymers and New Polymeric Materials?

The production of new polymers is the driving force that continues to replace metals and ceramics by synthetic polymers, and fuels the expansion of polymeric materials into fields as diverse as fuel cells, batteries, light-emitting devices, biomedical materials, and components of aircraft and automobiles. New polymeric materials arise through a combination of molecular design and the development of new polymer synthesis techniques. The principles about the relationship between macromolecular structure and properties form the basis of macromolecular design, and this in turn stimulates the development of new synthesis methods and new ways to tune the solid state and surface properties of polymers. In this sense, modern polymer chemistry, with its broad portfolio of known structure-property relationships, has an advantage over the earlier approach in which the (sometimes accidental) synthesis of a new polymer often yielded unexpected combinations of properties. However, it should be recognized that the element of surprise still exists in scientific work. Indeed, it is sometimes said that the role of the scientist is to do things that can not be predicted, and cannot always be justified on the basis of common knowledge. Only in this way can science escape from accepted practice and generate real breakthroughs into new areas. Work that follows an absolute adherence to well established knowledge and protocol may not yield any new insights.

Thus, useful new polymers are produced by three different routes—firstly, by a chance discovery; secondly, by a scientist who sees hitherto unrecognized connections between two different types of chemistry and combines them to produce a new system; and thirdly, by logical methodical design based on structure-property relationship. The third method has to be preceded by one of the first two. Developments that result from combining ideas from two different areas are often based on a subconscious knowledge of the general relationships between structure and properties in other systems.

The discovery of polydimethylsiloxane and polytetrafluoroethylene (Teflon), and the synthesis of high-density polyethylene arose from accidental reactions (by Kipping, Plunkett and Ziegler, respectively). The polyorganophosphazenes arose through an integration of ideas from organic polymer chemistry and inorganic chemistry.

Usually, accidental discoveries or cross-disciplinary discoveries of this type stimulate a burst of exploratory research which then leads to the synthesis of many different derivatives and the development of structure-property relationships, sometimes by researchers other than the discoverer. By some accounts, Kipping actually threw away the first polyorganosiloxanes made in his laboratory because they were not the compounds he was looking for. It was left to others to recognize their value and develop them in detail. There is a lesson here. On the other hand, most of the discoverers of new polymers are instantly aware of the significance of the discovery and set to work, sometimes in spite of widespread scepticism and opposition, to develop the field and establish the primary structure-property relationships needed to enable the new field to develop. After that, useful polymers are usually produced by direct design.

The developmental design approach often starts from a knowledge of the defects of existing polymeric materials, information that comes from engineers, physicists, physicians, and other users. Thus, the design process starts with a knowledge of the combination of properties needed to make a better material, and then proceeds to consider the alternative ways in which these properties can be generated. For example, if there is a need for fire-resistant clothing for firefighters or race-car drivers, this would require a polymer with a high melting point, and this in turn suggests the use of polymers with relatively stiff structure, Such as the aromatic polyamides. The key structure-property relationship involved here would be chain stiffness to give a high melting point. The choice of this type of polymer might then require exploratory synthesis work using condensation polymerization. Other decisions would be needed about how to generate crystallinity, in fibers of the polymer, because crystallinity would yield high tensile strength. Crystallinity would depend on monomer sequencing and on the types of side groups present. Resistance to fire might be achieved by the incorporation of phosphorus into the molecular structure, and so on.

Polymer chemists spend a great deal of their time designing new molecules and pondering their structure-property relationships, and the readers is encouraged to practice this procss routinely to gain experience.

——Allock H R, Laampe F W, Mark J E. Contemporary Polymer Chemistry, New York: Prentice Hall, INC. 2003

Words and Expressions

ceramics	[si'ræmiks]	n.	陶瓷制品
fuel	['fjuəl]	v.	推动,促进,加油;
		n.	燃料
fuel cell			燃料电池
diverse	[dai'və:s]	a.	多种多样的
battery	['bætəri]	n.	电池,蓄电池
biomedical material			生物医学材料
light-emitting device			发光器件
insight	['insait]	n.	见解,深刻的理解
chance discovery			偶然发现
hitherto	['hiðə'tu:]	adv.	迄今,至今
subconscious	['sʌb'kɔnʃəs]	a.	下意识的,潜意识的
polydimethylsiloxane	[pɔlidai'meθil'silikən]	n.	聚二甲基硅氧烷
polytetrafluoroethylene	[pɔlitetrə'fluərəu'eθili:n]	n.	聚四氟乙烯
component of aircraft			飞机零部件
automobiles	['ɔ:təməubi:l]	n.	汽车
stimulate	['stimjuleit]	v.	激发,促进
tune	[tju:n]	v.	调节,调整
in turn			反过来
portfolio	[pɔ:t'fəuljəu]	n.	文献资料
element of surprise			偶然因素
breakthrough	['breikθru:]	n.	突破性进展
accepted practice			习以为常的做法
absolute adherence to			严格遵守
protocol	['prəutəkɔl]	n.	惯例,协议
polyorganophosphazene	[pɔliɔ:'gænəu'fɔsfeizi:n]	n.	聚有机磷腈
integration	[,inti'greiʃən]	n.	结合
cross-disciplinary			交叉学科的
physicist	['fizisist]	n.	物理学家
physician	[fi'ziʃən]	n.	医生
exploratory	[iks'plɔ:rətəri]	a.	探索性的
ponder	['pɔndə]	v.	考虑,认真思考
scepticism	['skeptisizəm]	n.	怀疑
sequencing	['si:kwənsiŋ]	n.	序列
routinely	[ru:'ti:nli]	adv.	通常地

UNIT 6 Molecular Weight and Its Distributions of Polymers

The molecular weight of a polymer is of prime importance in its synthesis and application. The interesting and useful mechanical properties which are uniquely associated with polymeric materials are a consequence of their high molecular weight. Most important mechanical properties depend on and vary considerably with molecular weight. Thus, strength of polymer does not begin to develop until a minimum molecular weight of about 5000~10000 is achieved.① Above that size, there is a rapid increase in the mechanical performance of polymers as their molecular weight increases; the effect levels off at still higher molecular weights. In most instances, there is some molecular weight range in which a given polymer property will be optimum for a particular application.② The control of molecular weight is essential for the practical application of a polymerization process.

When one speaks of the molecular weight of a polymer, one means something quite different from that which applies to small-sized compounds.③ Polymers differ from the small-sized compounds in that they are polydisperse or heterogeneous in molecular weight. Even if a polymer is synthesized free from contaminants and impurities, it is still not a pure substance in the usually accepted sense. Polymers, in their purest form, are mixtures of molecules of different molecular weights. The reason for the polydispersity of polymers lies in the statistical variations present in the polymerization processes. When one discusses the molecular weight of a polymer, one is actually involved with its average molecular weight. Both the average molecular weight and the exact distribution of different molecular weights within a polymer are required in order to fully characterize it. The control of molecular weight and molecular weight distribution (MWD) is often used to obtain and improve certain desired physical properties in a polymer product.

Various methods are available for the experimental measurement of the average molecular weight of a polymer sample. These include methods based on colligative properties, light scattering, viscosity, ultracentrifugation, and sedimentation. The various methods do not yield the same average molecular weight. Different average molecular weights are obtained because the properties being measured are biased differently toward the different sized polymer molecules in a polymer sample.④ Some methods are biased toward the larger sized polymer molecules, while other methods are biased toward the smaller sized molecules. The result is that the average molecular weights obtained are correspondingly biased toward the larger or smaller sized molecules. The most important average molecular weights which are determined are the *number-average molecular weight* $\overline{M_n}$, the *weight-average molecular weight* $\overline{M_w}$ and the *viscosity-average molecular weight* $\overline{M_v}$.

In addition to the different average molecular weights of a polymer sample, it is frequently desirable and necessary to know the exact distribution of molecular weights. A variety of different fractionation methods are used to determine the molecular weight distribution of a polymer sample. These are based on fractionation of a polymer sample using properties, such as solubility and permeability, which vary with molecular weight.

——Odian G. Principles of Polymerization, New York: McGraw-Hill Book Company, 1973. 19

Words and Expressions

molecular weight distribution		n.	分子量分布
mechanical property			力学性能
minimum	[ˈminiməm]	n.	最小值
		a.	最小的
performance	[pəˈfɔːməns]	n.	性能，特征
strength	[streŋθ]	n.	强度
optimum	[ˈɔptiməm]	a.	最佳［优］的
		n.	最佳［优］值［点、状态］
polydisperse	[pɔlidisˈpəːs]	a.	多分散的
heterogeneous	[hetərəuˈdʒiːnjəs]	a.	不均匀的，非均相的
contaminant	[kɔnˈtæminənt]	n.	污物
polydispersity	[pɔlidisˈpəːsiti]	n.	多分散性
statistical	[stəˈtistikəl]	a.	统计的
variation	[vɛəriˈeiʃən]	n.	变化，改变
improve	[imˈpruːv]	v.	增进，改善
colligative	[ˈkɔligeitiv]	a.	依数的
light scattering			光散射
ultracentrifugation	[ʌltrəsentrifjuːˈgeiʃən]	n.	超速离心（分离）
sedimentation	[sedimenˈteiʃən]	n.	沉降（法）
number average molecular weight			数均分子量
weight average molecular weight			重均分子量
viscosity average molecular weight			黏均分子量
fractionation	[frækʃəˈneiʃən]	n.	分级
solubility	[sɔljuˈbiliti]		溶解度
permeability	[pəːmiəˈbiliti]	n.	渗透性

Phrases

be associated with… 与…有关
level off… 达到平衡，变平，趋缓
in most instances 在大多数情况下
be essential for… 对…是必需的
speak of… 谈到

free from… 没有…，无…
in the usually accepted sense 在能被人们广泛接受的意义上
be biased toward (s)… 有…偏向［差］，偏于…
a variety of… 各种各样的

Notes

① "Thus, strength of polymer does not begin to develop until a minimum molecular weight of about 5000~10000 is achieved." 该句的主句中 "develop" 意为 "显示出来"。主句后为 "until" 引导的时间状语从句。全句可直译为："因此，直到最小相对分子质量增大到大约 5000~10000 以前，聚合物的强度没有开始显示出来。"考虑到中文的习惯，可以译为："因此，直到最小相对分子质量增大到大约 5000~10000以后，聚合物的强度才开始显示出来。"

② "In most instances, there is some molecular weight range in which a given polymer property will be optimum for a particular application." 句中 "some" 作 "某一" 解。主句后为 "molecular weight range" 的定语从句。全句译文："在大多数情况下，对于某特定的应用来说，某种聚合物存在着某一分子量范围，在这个范围之内其性能是最好的。"

③ "When one speaks of the molecular weight of a polymer, one means something quite different from that which applies to small-sized compounds." 句子开头是一个时间状语从句，"one" 作 "人们" 解。在主句

中"different from that…"是"something"的后置定语，其中"that"代表前面的单数名词"something"。最后为"which"引导的代词"that"的定语从句。全句译文："当人们谈到聚合物分子量的时候，他所指的是和（适用于）低分子化合物的分子量完全不同的另一回事。"

④ "Different average molecular weights are obtained because the properties being measured are biased differently toward the different sized polymer molecules in a polymer sample." 主句后为原因状语从句，其中分词短语"being measured"为"properties"的后置定语。"are biased differently toward…"意指"对…有不同的偏差"。全句译文："（对同一聚合物）得到了不同的平均分子量，因为所测得的性质对试样中不同尺寸的聚合物分子有不同的偏差。"

Exercises

1. Choose the correct answer for each sentence
   (1) The properties of polyoxymethylene is similar _____ those of 6/6 Nylon.
       A. for   B. as   C. to   D. with
   (2) That resin can be used to produce plastic articles and ____ coatings.
       A. for   B. as   C. to   D. with
   (3) Carbonyl group represents ____ .
       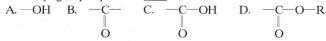
   (4) There are ____ 3000 kinds of electrical and electronic equipment at the exhibition.
       A. much   B. many   C. more   D. some
   (5) The ratio between saturated and unsaturated acid components in unsaturated polyesters will determine the degree of _____ .
       A. stereoregularity   B. crystallinity   C. crosslinking   D. branching
   (6) All metals are solid _____ mercury.
       A. in addition to   B. as well as   C. apart from   D. besides

2. Put the following words into Chinese
   polyvinyl alcohol   acetal   polyethylene oxide   polyoxymethylene   polyisobutylene   polyurethane
   epoxy resin   polyether   polyamide   polyvinylchloride

3. Put the following words into English
   离子型聚合   阴离子型聚合   阳离子型聚合   逐步聚合   加成聚合   自由基型聚合   共聚合   均聚
   缩聚   连锁聚合

Reading Materials

Purification of Polymers

The usual purification operations used with low molecular weight compounds (for example, distillation and recrystallization) are usually not applicable to macromolecular compounds.

Macromolecular compounds are not volatile and cannot simply be recrystallized from saturated solutions. One therefore has to be satisfied with extracting them in suitable solvents and thus removing the impurities.

The purification effect of the extraction is in most cases rather slight because in many instances the impurity is held to the polymer by strong secondary valence forces. Consequently, one usually first dissolves the macromolecular compound and then precipitates it by addition of a nonsolvent. The precipitate comes out as a more or less strongly swollen gel, and the impurities remain at least partly in solution. The gel can be removed by sedimentation if necessary, in the centrifuge. Usually it is necessary to repeat this precipitation 10 to 20 times. The solvents and precipitating agents have to be carefully selected from case to case. Many vinyl polymers are easily dissolved in benzene and toluene and precipitate under the addition of methanol. Methanol is a sort of general precipitating agent in which most polymers, both the benzene-soluble ones and the water-soluble polymers, are insoluble. Since methanol is miscible in all proportions both with benzene and with most other organic solvents and also with water, it can be used as a precipitant in both cases, i. e., with organic and with aqueous polymer solutions.

One has to expect that the polymer is somehow affected by the extraction, regardless of whether this is done by heating or by precipitation. Thus if one extracts with acids or bases, hydrolytic degradation usually causes a lowering in the degree of polymerization (for example, hydrolytic degradation of polysaccharides). In the process of dissolving and reprecipitation, one usually finds that low molecular weight components remain in solution (fractionation), and therefore the precipitated part has a higher average molecular weight and a more uniform molecular weight distribution. This may or may not be desirable, but in any case one should be aware of it.

With synthetic polymers, purification may be avoided sometimes by choosing a method of synthesis where the polymer is immediately obtained in pure form. Usually an extreme purification of the monomer is simpler and more effective than a later purification of the polymer. Initiators and other additives, such as emulsifying agents and protective colloids, are also selected wherever possible in such a way that they can remain in the polymer without causing problems.

Dispersions, as obtained by emulsion polymerization, can be coagulated or precipitated by the addition of electrolytes. Formic acid or sodium chloride is used most often. The purification effectiveness of such a precipitation is rather incomplete, because the emulsifying agents are not completely removed; their content is only lowered. However, even this is often sufficient to make precipitation of the polymer dispersion preferable to a direct drying on the mill or by spraying. Polymer dispersions can also often be precipitated and purified by pouring them into methanol, by adding methanol, or by cooling the dispersion in some kind of a cooling mixture ($-20°C$ to $-40°C$).

Of special importance is the purification of natural macromolecular compounds. These usually contain low molecular weight materials or other macromolecular compounds, and in some cases there is even a chemical combination between the different compounds: for example, cellulose with lignin (in wood), or pectins (in flax). In such cases it is justified to consider this combination as a new macromolecular substance. Since it is usually not possible to isolate the macromolecules without chemical reaction and therefore usually one does not obtain them in an unchanged form, one does not give these compounds a proper name but calls them protopectin or protocellulose, etc. The isolation of the pure compounds (pectin or cellulose) can be brought about by acid, or alkaline, or enzymatic hydrolysis (wood pulp by the sulfite process or by alkali treatment).

Low molecular weight impurities (for example, residues of solvents or monomers) can usually be removed by heating in vacuum. Since one cannot mix the molten polymer in some kind of a stirred vessel, (because of its high viscosity) one has to use special techniques and apparatuses, depending on the polymer. Macromolecular substances with a high softening point (over 100℃) can often be dried in the form of a powder or in granules. This is done by spreading them out in a vacuum oven and heating them, usually with an infrared lamp. If the material which is to be removed is volatile only at a temperature which is higher than the softening point of the polymer, one can use certain extruders in which it is possible to apply vacuum at certain places along the machine and thereby remove the solvent. In such vacuum extruders (welding machine) one can treat thermoplastic materials, such as polystyrene or polymethylmethacrylate (at 180℃ to 250℃). As a result of constant renewing of surface by the milling process in the vacuum-zones of the extruder, a very rapid and effective removal of the volatile compounds is obtained. Great care must be taken in the isolation of proteins from their aqueous solutions because they usually degenerate even at rather low temperatures. Such solutions can often be concentrated by freeze-drying. This is done by bringing the surface of the solution in close contact with a cold condenser and applying high vacuum to the entire apparatus so that the water from the solution forms as ice on the condenser. The solution to be concentrated can be in a frozen state during this process.

Low molecular weight impurities which are not volatile, such as inorganic salts in natural products, can be removed by dialysis, or electrodialysis, of the polymer solution.

The separation of different macromolecular compounds which are present in the mixture is particularly difficult. This is especially so if the solubility is the same, so that extraction or precipitation does not bring about a separation. This problem is often encountered in protein chemistry. Since the protein molecules are usually present in the solution in the form of ions, one can use the difference in their rate of diffusion in an electric field, (which depends on the number of acid and basic amino acids in the macromolecule), for the separation of different proteins. This process is called electrophoresis. For an electrophoretic separation, one needs a special apparatus which permits optical control and registration of the separation process, and which in many cases also permits a preparative separation of the components.

Another process, gel permeation chromatography (GPC), has been used with great success for the purification of polymers, especially for the separation of different polypeptides. Because of its general importance for the fractionation of polymers, together with the other methods of fractionation.

——Vollmert B. Polymer Chemistry. Berlin: Springer-Verlag, 1973. 305

Words and Expressions

purification	[pjuərifi'keiʃən]	n.	提纯，净化
extract	[iks'trækt]	v.	萃取
nonsolvent	[nɔn'sɔlvənt]	n.	非溶剂
miscible	['misibl]	a.	可混溶的
precipitant	[pre'sipitənt]	n.	沉淀剂
hydrolytic	[hai'drɔlitik]	a.	水解的
polysaccharide	[pɔli'sækəraid]	n.	多糖，聚糖
protective colloid			保护胶体
electrolyte	[i'lektrəulait]	n.	电解质
formic acid			甲酸，蚁酸
cellulose	['seljuləus]	n.	纤维素
lignin	['lignin]	n.	木质素
pectin	['pektin]	n.	果胶

flax	[flæks]	n.	亚麻纤维
isolate	['aisəleit]	v.	隔离，析出，绝缘
protopectin	[prəutəu'pektin]	n.	原果胶
enzymatic hydrolysis			酶催化水解
residue	['rezidju:]	n.	残余物
vessel	['vesl]	n.	容器
degenerate	[di'dʒenəreit]	v.	退化，变性，变质
freeze-drying			冷冻干燥
frozen state			冷冻［冻结］状态
dialysis	['daiəlaisis]	n.	透析，渗析
electrodialysis	[i'lektrəudai'ælisis]	n.	电渗析
protein	['prəuti:n]	n.	蛋白质
electrophoresis	[i'lektrəufə'ri:sis]	n.	电泳（现象）
polypeptide	[pɔli'peptaid]	n.	多肽

UNIT 7 Polymer Solution

Dissolving a polymer is a slow process that occurs in two stages. First, solvent molecules slowly diffuse into the polymer to produce a swollen gel. This may be all that happens if, for example, the polymer-polymer intermolecular forces are high because of crosslinking, crystallinity, or strong hydrogen bonding.① But if these forces can be overcome by the introduction of strong polymer-solvent interactions, the second stage of solution can take place. Here the gel gradually disintegrates into a true solution. Only this stage can be materially speeded by agitation. Even so, the solution process can be quite slow (days or weeks) for materials of very high molecular weight.

Solubility relations in polymer systems are more complex than those among low-molecular-weight compounds, because of the size differences between polymer and solvent molecules, the viscosity of the system, and the effects of the texture and molecular weight of the polymer. In turn, the presence or absence of solubility as conditions (such as the nature of the solvent, or the temperature) are varied can give much information about the polymer.②

As specified in the literature, the arrangements of the polymer chain differing by reason of rotations about single bonds are termed *conformations*.③ In solution, a polymer molecule is a randomly coiling mass most of whose conformations occupy many times the volume of its segments alone. The average density of segments within a dissolved polymer molecule is of the order of $10^{-4} \sim 10^{-5} \text{g/cm}^3$. The size of the molecular coil is very much influenced by the polymer-solvent interaction forces. In a thermodynamically "good" solvent, where polymer-solvent contacts are highly favored, the coils are relatively extended. In a "poor" solvent they are relatively contracted. It is the purpose to describe the conformational properties of both ideal and real polymer chains.

The importance of the random-coil nature of the dissolved, molten, amorphous, and glassy states of high polymers cannot be overemphasized. Many important physical as well as thermodynamic properties of high polymers result from this characteristic structural feature. The random coil (Fig. 7.1) arises from the relative freedom of rotation associated with the chain bonds of most polymers and the formidably large number of conformations accessible to the molecule.④

Fig. 7.1

Model of one of many conformations of a random-coil chain of 1000 links

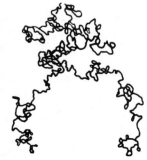

One of these conformations, the fully extended chain has special interest because its length, the *contour length* of the chain, can be calculated in a straightforward way. In all other cases the size of the random coil must be expressed in terms of statistical parameters such as the root-mean-square distance between its ends, $(\overline{r^2})^{1/2}$, or its *radius of gyration*, the root-mean-square distance of the elements of the chain from its center of gravity, $(\overline{s^2})^{1/2}$. For linear polymers that are not appreciably extended beyond their most probable shape, the mean-square end-to-end distance and the mean-square of the radius of gyration are simply related: $\overline{r^2} = 6\overline{s^2}$. For extended chains $\overline{r^2} > 6\overline{s^2}$. The use of the radius of gyration is sometimes preferred because it can be determined experimentally.

——Fred W, Billmeyer J. Textbook of Polymer Science, New York: John Wiley & Sons, 1984. 151

Words and Expressions

diffuse	[diˈfjuːz]	v.	扩散
swollen	[ˈswəulən]	a.	溶胀的
gel	[dʒel]	n.	凝胶
intermolecular	[ˌintəməˈlekjulə]	a.	(作用于) 分子间的
crosslinking	[ˈkrɔːsliŋkiŋ]	n.	交联
crystallinity	[ˌkristəˈliniti]	n.	结晶性, 结晶度
hydrogen bonding			氢键
disintegrate	[diˈsintigreit]	v.	分解, 分散, 分离
interaction	[ˌintəˈrækʃən]	n.	相互作用
agitation	[ˌædʒiˈteiʃən]	n.	搅拌
texture	[ˈtekstjə]	n.	结构, 组织
arrangement	[əˈreindʒmənt]	n.	(空间) 排布, 排列
single bond			单键
conformation	[ˌkɔnfɔːˈmeiʃən]	n.	构象
coiling	[ˈkɔiliŋ]	a.	线团状的
density	[ˈdensiti]	n.	密度
segment	[ˈsegmənt]	n.	链段
coil	[ˈkɔil]	n.	线团
thermodynamically	[ˌθəːməudaiˈnæmikəli]	adv.	热力学地
ideal	[aiˈdiəl]	a.	理想的, 概念的
real	[ˈriəl]	a.	真实的
random coil			无规线团
molten	[ˈməultən]	a.	熔化 [融] 的
amorphous	[əˈmɔːfəs]	a.	无定形的, 非晶态的
glassy	[ˈglɑːsi]	a.	玻璃 (态) 的
contour	[ˈkɔntuə]	n.	外形, 轮廓
radius	[ˈreidjəs]	n.	半径
gyration	[dʒaiəˈreiʃən]	n.	旋转, 回旋
mean-square end-to-end distance			均方末端距
model	[ˈmɔdl]	n.	模型

Phrases

even so　虽然如此
in turn　(本身) 又 [也]; 依次
by reason of...　由于, 因为
arise from　由于…而产生 [造成] 的, 起因于…

accessible to...　为…所能达到的
in a straightforward way　直接地
in terms of...　根据, 借助于, 利用, 就…而论

Notes

① "This may be all that happens if, for example, the polymer-polymer intermolecular forces are high because of crosslinking, crystallinity, or strong hydrogen bonding." 主句为 "This may be all", 其中 "This" 为主语, 代表前句中所指 "生成被溶剂溶胀的凝胶 (swollen gel)" 这一阶段; "all" 为名词, 作

主句的表语，其后的"that happens"为"all"的定语从句，"This may be all that happens"可直译为"这可能就是所发生的一切了"，这不符合中文习惯，意译为"有可能就只停留在（生成被溶剂溶胀的凝胶）这一阶段"。"if"之后为主句的条件状语从句，"because of"引导的介词短语作条件从句的原因状语。全句译文："例如，如果因交联、结晶和很强的氢键而形成很大的分子间力，（聚合物的溶解过程）有可能就只停留在这一阶段。"

② "In turn, the presence or absence of solubility as conditions (such as the nature of the solvent, or the temperature) are varied can give much information about the polymer."句中"in turn"作"（本身）又"解。"the presence or absence of solubility"可译作"有无溶解性"，作主句的主语。"can give"为主句的谓语。"much information about the polymer"为主句的宾语，意指"关于聚合物（内部结构、组成、分子量等）的许多信息"。在主句中插入一个由"as"引导的时间状语从句。全句译文："当条件（溶剂的本质或温度）变化的时候，有无溶解性又可以提供出许多关于这种聚合物的信息。"

③ "As specified in the literature, the arrangements of the polymer chain differing by reason of rotations about single bonds are termed *conformations*."句首为由"As"引导的定语从句，说明整个主句，"As"在从句中作主语。主句中分词"differing"为主语"the arrangements"的后置定语。"by reason of rotation about single bonds"为"differing"的原因状语，其中"about"作"围绕"解。全句译文："正如在文献中所定义的那样，由于围绕着单键的旋转而导致的聚合物链不同的空间排布叫作构象。"

④ "The random coil (Fig. 7.1) arises from the relative freedom of rotation associated with the chain bonds of most polymers and the formidably large number of conformations accessible to the molecule."该句子为一个简单句，谓语"arise from"可译为"是由…而产生的"。第一个宾语"the relative freedom of rotation"意指"相当自由的旋转"。"associated with…"为"freedom"的后置定语。第二个宾语"the formidably large number"意为"巨大的数目"，"accessible to…"为"number"的后置定语。全句译文："无规线团（图7.1）一方面是由于聚合物链上的键自由旋转而产生的，另一方面是由于（聚合物）分子（链）可达到巨大的构象数而产生的。"

Exercises

1. *Translate the following passage into Chinese*

   Ring-opening polymerizations proceed only by ionic mechanisms, the polymerization of cyclic ethers mainly by cationic mechanisms, and the polymerization of lactones and lactams by either a cationic or anionic mechanism. Important initiators for cyclic ethers and lactone polymerization are those derived from aluminum alkyl and zinc alkyl/water systems. It should be pointed out that substitution near the reactive group of the monomer is essential for the individual mechanism that operates effectively in specific cases; for example, epoxides polymerize readily with cationic and anionic initiators, while fluorocarbon epoxides polymerize exclusively by anionic mechanisms.

2. *Write out an abstract in Chinese for the text in this unit*
3. *Please fill in the correct answers at the blanks in the following table*

Chinese	English	molecular structure
	polybutadiene	
聚苯乙烯		
		Cl—Ti—Cl | Cl
四氢呋喃		
		$CH_3CH_2CH_2CH_2$—Li
	isobutylene	

续表

Chinese	English	molecular structure
酚钠		
		NaCl
		CH$_2$=CH—CH$_2$—
	phosgene	

Reading Materials

Viscous Flow

1. Phenomena of Viscous Flow

If a force per unit area s causes a layer of liquid at a distance x from a fixed boundary wall to move with a velocity ν, the viscosity η is defined as the ratio between the shear stress s and the velocity gradient $\partial \nu / \partial x$ or rate of shear γ:

$$s = \eta \frac{\partial \nu}{\partial x} = \eta \gamma \tag{7.1}$$

If η is independent of the rate of shear, the liquid is said to be *Newtonian* or to exhibit ideal flow behavior (Fig. 7.2 (a)). Two types of deviation from Newtonian flow are commonly observed in polymer solutions and melts. One is *shear thinning* or *pseudoplastic* behavior, a reversible decrease in viscosity with increasing shear rate (Fig. 7.2 (b)). Shear thinning results from the tendency of the applied force to disturb the long chains from their favored equilibrium conformation, causing elongation in the direction of shear. An opposite effect, *shear thickening or dilatant* behavior (Fig 7.2 (c)), in which viscosity increases with increasing shear rate, is not observed in polymer.

A second deviation from Newtonian flow is the exhibition of a *yield value*, a critical stress below which no flow occurs. Above the yield value, flow may be either Newtonian (as indicated in Fig. 7.2 (d)) or non-Newtonian. For most polymer melts, only an apparent yield value is observed.

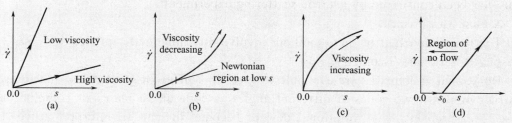

Fig. 7.2

Dependence of shear rate $\dot{\gamma}$ on shear stress s
(a) Newtonian; (b) pseudoplastic; (c) dilatant behavior;
(d) the presence of a yield stress s_0 followed by Newtonian behavior

The above effects are shear dependent but time independent. Some fluids also exhibit reversible time-dependent changes in viscosity when sheared at constant stress. Viscosity decreases with time in a *thixotropic fluid*, and increases with time in a *rheopectic* fluid, under constant shear stress.

For low-molecular-weight liquids, the temperature dependence of viscosity is found to

follow the simple exponential relationship:
$$\eta = A e^{E/RT} \tag{7.2}$$
where E is an *activation energy for viscous flow* and A is a constant.

These features of the flow of liquids can be explained in terms of several molecular theories. That of Eyring is based upon a lattice structure for the liquid, containing some unoccupied sites or holes. These sites move at random throughout the liquid as they are filled and created anew by molecules jumping from one site to another. Under an applied stress the probability of such jumps is higher in the direction that relieves the stress. If each jump is made by overcoming an energy barrier of height E, the theory leads to Eq. (7.2). The energy of activation E is expected to be related to the latent heat of vaporization of the liquid, since the removal of a molecule from the surroundings of its neighbors forms a part of both processes. Such a relation is indeed found and is taken as evidence that the particle that moves from site to site is probably a single molecule.

As molecular weight is increased in a homologous series of liquids up to the polymer range, the activation energy of flow E does not increase proportionally with the heat of vaporization but levels off at a value independent of molecular weight. This is taken to mean that in long chains the unit of flow is considerably smaller than the complete molecule. It is rather a segment of the molecule whose size is of the order of 5~50 carbon atoms. Viscous flow takes place by successive jumps of segments (with, of course, some degree of coordination) until the whole chain has shifted.

2. Dynamics of Polymer Melts

It is now accepted that polymer chains are strongly intertwined and entangled in the melt; the dynamic behavior of such a system has been reviewed, but is only poorly understood. Thermodynamically, the chains are essentially ideal, as was first realized by Flory. Their freedom of motion results from the presence of a *correlation hole* around each flow unit, within which the concentration of similar units from other chains is reduced. The presence of these correlation holes, and of the ideal but entangled nature of chains in the melt, has been confirmed by neutron-scattering experiments.

3. Flow Measurement

The most important of these methods involve rotational and capillary devices.

(1) Rotational Viscometry

Rotational viscometers are available with several different geometries, including concentric cylinders, two cones of different angles, a cone and a plate, or combinations of these. Measurements with rotational devices become difficult to interpret at very high shear stress, owing to the generation of heat in the specimen because of dissipation of energy, and to the tendency of the specimen to migrate out of the region of high shear. This phenomenon, the *Weissenberg effect*, arises because the stress in any material can always be analyzed into the components of a 3×3 stress tensor, in which the off-diagonal elements, called *normal stresses* because they act perpendicular to the surface of the specimen, are not negligible in viscoelastic fluids.

A simple rotational instrument used in the rubber industry is the *Mooney viscosimeter*. This empirical instrument measures the torque required to revolve a rotor at constant speed in a sample of the polymer at constant temperature. It is used to study changes in the flow characteristics of rubber during milling or mastication. The *Brabender Plastograph* is a similar device.

(2) Capillary Viscometry

Capillary rheometers, usually made of metal and operated either by dead weight or by gas pressure, or at constant displacement rate, have advantages of good precision, ruggedness, and ease of operation. They may be built to cover the range of shear stresses found in commercial fabrication operations. However, they have the disadvantage that the shear stress in the capillary varies from zero at the center to a maximum at the wall.

An elementary capillary rheometer (extrusion plastometer) is used to determine the flow rate of polyethylene in terms of *melt index*, defined as the mass rate of flow of polymer through a specified capillary under controlled conditions of temperature and pressure.

——Fred W Billmeyer J. Textbook of Polymer Science. New York: John Wiley & Sons, 1984. 304

Words and Expressions

viscous flow			黏性流
shear stress			剪切应力
rate of shear			剪切速率
Newtonian fluid			牛顿型流体
pseudoplastic	['psju:dəuplæstik]	a.	假塑性的
dilatant	[dai'leitənt]	a.	胀塑[流]型的
thixotropic fluid			触变性流体
rheopectic fluid			震凝(性)流体
lattice	['lætis]	n.	晶格,点阵,网络结构
latent heat of vaporization			蒸发潜热
dynamics	[dai'næmiks]	n.	动力学,动态(特征)
rotational viscometer			旋转黏度计
viscoelastic fluid			黏弹性流体
capillary rheometer			毛细管流变仪
relaxation	[ri:læk'seiʃən]	n.	松弛
mastication	[mæsti'keiʃən]	n.	素炼
ruggedness	['rʌgidnis]	n.	坚固性,耐久性

UNIT 8 Morphology of Solid Polymers

Solid polymers differ from ordinary, low molecular weight compounds in the nature of their physical state or morphology. Most polymers simultaneously show the characteristics of both crystalline solids and highly viscous liquids. X-ray and electron diffraction patterns often show the sharp features typical of three-dimensionally ordered, crystalline materials as well as the diffuse features characteristic of liquids. ① The term *crystalline and amorphous* are usually used to indicate the ordered and unordered polymer regions, respectively. Different polymers show different degree of crystalline behavior. Although a few polymers may be completely amorphous and a few completely crystalline, most polymers are partially or *semi-crystalline* in character.

The exact nature of polymer crystallinity has been the subject of considerable controversy. The *fringed-micelle* theory, developed in the 1930's, considers polymers to consist of small-sized, ordered crystalline regions-termed crystallites-imbedded in an unordered, amorphous polymer matrix. Polymer molecules are considered to pass through several different crystalline regions with crystallites being formed when segments from different polymer chains are precisely aligned together and undergo crystallization. ② Each polymer chain can contribute ordered segments to several crystallites. The segments of the chain in between the crystallites make up the unordered amorphous matrix. This concept of polymer crystallinity is shown in Fig. 8.1.

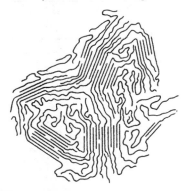

Fig. 8.1

Fringed-micelle picture of polymer crystallinity

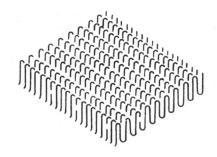

Fig. 8.2

Folded-chain picture of polymer crystallinity

The *folded-chain lamella* theory arose in the late 1950's when polymer single crystals in the form of thin platelets termed *lamella* were grown from polymer solutions. The diffraction patterns of these single crystals indicate that the polymer molecules fold back and forth on themselves like in an accordion in the process of crystallization (Fig. 8.2). The theory of chain-folding applies generally to most polymers—not only for solution-grown single crystals, but also for polymers crystallized from the melt. Semi-crystalline polymers are considered by advocates of the folded-chain theory to be chain-folded crystals with varying amounts of defects. ③ The crystallinity of polymers is pictured as being completely similar to that of low molecular weight compounds. The defects in the chain folded crystals may be imperfect folds, irregularities in packing, chain entanglements, loose chain ends, dislocations, occluded impurities, or numerous other imperfections.

The fringed-micelle and folded-chain theories of polymer crystallinity are often considered to be mutually exclusive but they need not be so considered. It is usually practical to adopt a working model of polymer crystallinity which employs the features of both concepts. The folded chain theory is especially well suited for highly crystalline polymers where one can consider them to be one phase crystalline systems with defects. Polymers with medium to low crystallinity can often be advantageously treated by the fringed-micelle concept as two-phase systems composed of crystallites imbedded in uncrystallized, amorphous polymer. ④ The structure of the crystallites in such polymers may be that of the folded-chain lamella. The extent and type of crystallinity in a polymer is experimentally determined by methods such as density, X-ray and electron diffraction, infrared spectroscopy, and nuclear magnetic resonance. The results are often interpreted in a simplified manner as a weight or volume percent crystallinity by comparison with measurements on completely crystalline and completely amorphous polymer samples.

——Odian G. Principles of Polymerization. New York: McGraw-Hill Book Company, 1970. 24

Words and Expressions

morphology	[mɔː'fɔlədʒi]	n.	形态（学）
simultaneously	[siməl'teinjəsli]	adv.	同时，同步
crystalline	['kristəlain]	n.	晶体，晶态
		a.	结晶的，晶态［状］的
diffraction	[di'frækʃən]	n.	衍射
pattern	['pætən]	n.	花纹，图样式样
X-ray		n.	X射线，X光
three dimensionally ordered			三维有序的
semicrystalline	[semi'kristəlain]	n.	半晶
controversy	[kɔntrə'vəːsi]	n.	争论，争议
fringed-micelle theory			缨状微束理论
crystallite	['kristəlait]	n.	微晶
imbed	[im'bed]	v.	嵌入，埋入，包埋
matrix	['meitriks]	n.	基体，母体，基质，矩阵
align	[ə'lain]	v.	排列成行
folded-chain lamella theory			折叠链片晶理论
platelet	['pleitlit]	n.	片晶
accordion	[ə'kɔːdiən]	n.	手风琴
defect	[di'fekt]	n.	缺陷
imperfect	[im'pəːfikt]	a.	不完全［整］的
irregularity	[iregju'læriti]	n.	不规则性，不均匀性
packing	['pækiŋ]	n.	堆砌
entanglement	[in'tæŋglmənt]	n.	缠结，纠缠
dislocation	[dislə'keiʃən]	n.	错位，位错
occluded	[ək'luːdid]	a.	夹杂［带］的
medium	['miːdiəm]	n.	介质
		a.	中等的，中间的
infrared spectroscopy		n.	红外光谱法
nuclear magnetic resonance		n.	核磁共振

Phrases

in the nature of... 按…的本性

(be) typical of... 是…的特征，具有…的特征，象征者

in character 如所预料，与本身特性相符
consider...to do 把…看作，[认为]是干…的
pass through 通过
contribute...to 把…贡献给…
in between 在…之间的，位于其间的

in the form of... 以 [取] …形态 [形状]，呈…状态
back and forth 来回，前后
be pictured as... 可想象为…
treat...as... 把… 作为…来处理

Notes

① "X-ray and electron diffraction patterns often show the sharp features typical of three-dimensionally ordered, crystalline materials as well as the diffuse features characteristic of liquids." 这句话是一个简单句，有两个并列的宾语，第一个宾语为 "sharp features"，紧连在其后的分词短语 "（being）typical of..." 为第一宾语的后置定语，在该分词短语语中省略了 "being"；第二个宾语为 "diffuse features"，紧连在其后的分词短语 "（being）characteristic of..." 为第二宾语的后置定语，在该分词短语中也省略了 "being"。本句译文："（固体聚合物的）X射线图样和电子衍射图样常常会显示出三维有序晶体材料的图样所具有的边缘清晰的特征，也会显示出液体的图样所具有的边缘模糊的特征。"

② "Polymer molecules are considered to pass through several different crystalline regions with crystallites being formed when segments from different polymer chains are precisely aligned together and undergo crystallization." 句中 "with crystallites being formed" 为介词后的主谓结构，作主句的伴随状语。其后为 "when" 引导的时间状语从句。本句译文："当来自不同聚合物链的链段精确地排列在一起结晶时，认为伴随着微晶的生成，聚合物分子会通过几个不同的晶区。"

③ "Semi-crystalline polymers are considered by advocates of the folded-chain theory to be chain-folded crystals with varying amounts of defects." 句中介词短语 "with varying amount of defects" 是 "crystals" 的后置定语。本句译文："折叠链理论的支持者们认为半晶聚合物为具有不同数量缺陷的折叠链晶体。"

④ "Polymers with medium to low crystallinity can often be advantageously treated by the fringed-micelle concept as two-phase systems composed of crystallites imbedded in uncrystallized, amorphous polymer." 句中介词短语 "with medium to low crystallinity" 为主语 "polymers" 的后置定语。"composed of" 引导的过去分词短语为 "two-phase systems" 的后置定语。"imbedded in" 引导的过去分词短语为 "crystallites" 的后置定语。本句译文："具有从中等到较低结晶度的聚合物常常适用于缨状微束理论作为两相体系来处理，这些体系是由微晶包埋在非结晶的无定形聚合物中而构成的。"

Exercises

1. *Please fill in the correct answers into the blanks in the following passage*

Polymers can be classified into two main groups, addition polymers and _____ polymers. This classification is based on whether or not the repeating unit of the polymer contains the same atoms _____ the monomer. The repeating unit of an addition polymer is identical _____ the monomer, while condensation polymers contain _____ because of the formation of _____ during the polymerization process. The corresponding polymerization processes would then be called addition polymerization and condensation polymerization. As was mentioned earlier, this classification can result _____ confusion, since it has been shown in later years that many important types of polymers can be _____ by both addition and condensation processes. For example, polyesters, polyamides and polyurethanes are usually considered to be _____ polymers, but they can be prepared by addition as well as by condensation reaction. Similarly, polyethylene normally considered an _____ polymer, can also be prepared by _____ reaction.

2. *Answer the following questions in English*

(1) What is chain polymerization?

(2) Which kinds of monomers can carry out step-growth polymerization process?

(3) What properties of polymers can be based on for measuring the molecular weight?

3. *Please write out at least* 10 *kinds of polymers both in English and in Chinese*
4. *Please write out at least* 10 *kinds of monomers both in English and in Chinese and the corresponding chemical structure*

Reading Materials

Physical Structure of Polymers

When the chemical composition of a polymer has been determined, there remains the important question of the arrangement of the molecular chains in space. This has two distinct aspects.

(1) The arrangement of a single chain without regard to its neighbors: rotational isomerism.

(2) The arrangement of chains with respect to each other: orientation and crystallinity.

1. Rotational Isomerism

The arrangement of a single chain relates to the fact that there are alternative conformations for the molecule because of the possibility of hundred rotation about the many single bonds in the structure. Rotational isomerism has been carefully studied in small molecules using spectroscopic techniques and the *trans* and *gauche* isomerism of disubstituted ethanes. Similar considerations apply to polymers.

To pass from one rotational isomeric form to another requires that an energy barrier be surmounted. The possibility of the chain molecules changing their conformations therefore depends on the relative magnitude of the energy barrier compared with thermal energies and the perturbing effects of applied stresses. We can therefore see the possibilities of linking molecular flexibility to deformation mechanisms.

2. Orientation and Crystallinity

When we consider the arrangement of molecular chains with respect to each other, there are again two largely separate aspects, those of molecular *orientation* and *crystallinity*. In semicrystalline polymers this distinction may at times be an artificial one.

Many polymers when cooled from the molten state form a disordered structure, which is termed the amorphous state. At room temperature these polymers may be of high modulus, such as polymethyl methacrylate, polystyrene and melt-quenched polyethylene terephthalate, or of low modulus, such as rubber or atactic polypropylene. Amorphous polymers are usually considered to be a random tangle of molecules. It is, however, apparent that completely random packing cannot occur. This follows even from simple arguments based on the comparatively high density but there is no distinct structure as, for example, revealed by X-ray diffraction techniques.

Polymethyl methacrylate, polystyrene and melt-quenched polyethylene terephthalate are examples of amorphous polymers. If such a polymer is stretched, the molecules may be preferentially aligned along the stretch direction. In polymethyl methacrylate and polystyrene such molecular orientation may be detected by optical measurements; but X-ray diffraction measurements still reveal no sign of three-dimensional order. The structure is therefore regarded as a somewhat elongated tangled skein. We would say that such a structure is oriented amorphous, but not crystalline.

In polyethylene terephthalate, however, stretching produces both molecular orientation and small regions of three-dimensional order, namely crystallites. The simplest explanation of such behavior is that the orientation processes have brought the molecules into adequate juxtaposition for them to take up positions of three-dimensional order, and hence crysatllize.

Many polymers, including polyethylene terephthalate, also crystallize if they are cooled slowly from the melt. In this case we may say that they are crystalline but oriented. Although such specimens are unoriented in the macroscopic sense; i. e. they possess isotropic bulk mechanical properties, they are not homogeneous in the microscopic sense and often show a spherulitic structure under a polarizing microscope.

The crystal structures of the crystalline regions can be determined from the wide angle X-ray diffraction patterns of polymers in the stretched crystalline form. Such structures have been obtained for all the well known crystalline polymers, e. g. polyethylene, nylon, polyethylene terephthalate, polypropylene, etc.

Such information is extremely valuable in gaining an understanding of the structure of a polymer. It was early recognized, however, that in addition to the discrete reflections from the crystallites the diffraction pattern of a semicrystalline polymer also shows much diffuse scattering and this was attributed to the "amorphous" regions.

This led to the so-called fringed micelle model for the structure of a semicrystalline polymer, which is a natural development of the imagined situation in an amorphous polymer. The molecular chains alternate between regions of order (the crystallites) and disorder (the amorphous regions).

This fringed micelle model was called into question by the discovery of polymer single crystals grown from solution. Linear polyethylene, when crystallized from dilute solution, forms single crystal lamellae, with lateral dimensions of the order of $10 \sim 20$ μm and of the order of 10 nm thick. Electron diffraction shows that the molecular chains are approximately normal to the lamellar surface and since the molecules are usually of the order of 1μm in length, it can be deduced that they must be folded back and forth within the crystals. The initial proposal suggested that the folds were sharp and regular (adjacent re-entry). Recent neutron scattering experiments suggest that the chain-folded ribbon itself doubles up beyond a certain length, which has been termed *superfolding* and that there is a definite departure from strict adjacent re-entry. This is an area of some controversy and one view is that the folds are predominantly non-adjacent. Single crystals have been isolated for most crystalline polymers, including nylon and polypropylene.

The influence of chain folding on the structure of bulk crystalline polymers is a matter of further controversy, although there is much evidence to support the existence of a lamellar morphology, and this is sometimes vital to the understanding of mechanical properties.

From a molecular point of view neutron diffraction experiments have again provided key imformation in showing that only small changes ocuur in the radius of gyration of the chain molecules in several polymers in crystallization from the molten state. We are bound to conclude that the highly entangled topology of the chains which exists in the melt must be substantially retained in the semicrystalline state. One model (the solidification model) proposes that crystallization occurs by straightening of coil sequences without involving long range diffusion processes.

Nevertheless, it is still necessary to understand the origins of the lamellar morphology of crystalline polymers. When a polymer melt is cooled, crystallization is initiated at nuclei in differnet points of the specimen. Spherulitic textures are then formed by the growth of dominant lamellae from a central nucleus in all directions by twisting of these lamellae along fibrils, the intervening spaces being filled in by subsidiary lamellae and possibly low molecular weight material.

The present state of knowledge suggests that it is most likely that in crystalline polymers chain folding occurs in addition to the more conventional threading of molecules through the crystalline regions.

——Ward I M. Mechanical Properties of Solid Polymers. Chichester: John Wiley & Sons, 1983.6

Words and Expressions

rotational isomerism		n.	旋转异构
orientation	[ɔːrienˈteiʃən]	n.	取向
hindered rotation			受阻旋转
spectroscopic technique			光谱技术
trans isomerism			反式异构
gauche isomerism			旁式异构
surmount	[səˈmaunt]	v.	克服，越过
perturbing effect			扰动效应
flexibility	[fleksəˈbiliti]	n.	柔性
disordered structure			无序结构
modulus	[ˈmɔdjuləs]		模量
polymethyl methacrylate			聚甲基丙烯酸甲酯
atactic	[əˈtæktik]	a.	无规立构的
polyethylene terephthalate			聚对苯二甲酸乙二醇酯，聚酯，涤纶
juxtaposition	[dʒʌkstəpəˈziʃən]	n.	并置，并列
isotropic	[aisəuˈtrɔpik]		各向同性的
spherulitic structure			球晶结构
polarizing microscope			偏光显微镜
fibril	[ˈfaibril]	n.	原纤，微纤

UNIT 9 Structure and Properties of Polymers

Most conveniently, polymers are generally subdivided in three categories, viz., plastics, rubbers and fibers. In terms of initial elastic modulus, rubbers ranging generally between 10^6 to 10^7 dynes/cm^2, represent the lower end of the scale, while fibers with high initial moduli of 10^{10} to 10^{11} dynes/cm^2 are situated on the upper end of the scale; plastics, having generally an initial elastic modulus of 10^8 to 10^9 dynes/cm^2, lie in-between. As is found in all phases of polymer chemistry, there are many exceptions to this categorization. ①

An elastomer (or rubber) results from a polymer having relatively weak interchain forces and high molecular weights. When the molecular chains are "straightened out" or stretched by a process of extension, they do not have sufficient attraction for each other to maintain the oriented state and will retract once the force is released. ② This is the basis of elastic behavior.

However, if the interchain forces are very great, a polymer will make a good fiber. Therefore, when the polymer is highly stretched, the oriented chain will come under the influence of the powerful attractive forces and will "crystallize" permanently in a more or less oriented matrix. These crystallization forces will then act virtually as crosslinks, resulting in a material of high tensile strength and high initial modulus, i.e., a fiber. Therefore, a potential fiber polymer will not become a fiber unless subjected to a "drawing" process, i.e., a process resulting in a high degree of intermolecular orientation. ③

Crosslinked species are found in all three categories and the process of crosslinking may change the cited characteristics of the categories. Thus, plastics are known to possess a marked range of deformability in the order of 100% to 200%; they do not exhibit this property when crosslinked, however. Rubber, on vulcanization, changes its properties from low modulus, low tensile strength, low hardness, and high elongation to high modulus, high tensile strength, high hardness, and low elongation. Thus, polymers may be classified as noncrosslinked and crosslinked, and this definition agrees generally with the subclassification in thermoplastic and thermoset polymers. From the mechanistic point of view, however, polymers are properly divided into addition polymers and condensation polymers. Both of these species are found in rubbers, plastics, and fibers.

In many cases polymers are considered from the mechanistic point of view. Also, the polymer will be named according to its source whenever it is derived from a specific hypothetical monomer, or when it is derived from two or more components which are built randomly into the polymer. This classification agrees well with the presently used general practice. When the repeating unit is composed of several monomeric components following each other in a regular fashion, the polymer is commonly named according to its structure.

It must be borne in mind that, with the advent of Ziegler-Natta mechanisms and new techniques to improve and extend crystallinity, and the closeness of packing of chains, many older data given should be critically considered in relation to the stereoregular and crystalline structure. ④ The properties of polymers are largely dependent on the type and extent of both stereoregularity and crystallinity. As an example, the densities and melting points of atactic and isotactic species are presented in Table 9.1.

Table 9.1 Densities and Melting Points of Atactic and Isotactic of Polyolefins

Project	Polypropylene		Polybutene	
	Atactic	Isotactic	Atactic	Isotactic
Densities/ (g/cm^3)	0.85	0.92	0.87	0.91
Melting points/℃	80	165	62	128

——Boenig H V. Structure and Properties of Polymers. Georg Thieme Publishers Stuttgart，1973. 18

Words and Expressions

subdivide	[sʌbdi'vaid]	v.	细分，区分
category	['kætigəri]	n.	种类，类型
plastics	['plæstiks]	n.	塑料
rubber	['rʌbə]	n.	橡胶
fiber	['faibə]	n.	纤维
elastic modulus			弹性模量
categorization	[kætigəri'zeiʃən]	n.	分类（法）
elastomer	[i:'læstəmə]	n.	弹性体
interchain	[intə'tʃein]	a.	链间的
stretch	[stretʃ]	v.	拉直，拉长
attraction	[ə'trækʃən]	n.	引力，吸引
orient	['ɔ:riənt]	v.	定向，取向
retract	[ri'trækt]	v.	收缩
release	[ri'li:s]	v.	解除，松开
tensile strength			抗张强度
deformability	[difɔ:mə'biliti]	n.	变形性，形变能力
vulcanization	[vʌlkənai'zeiʃən]	n.	硫化
hardness	['hɑ:dnis]	n.	硬度
elongation	[ilɔ:ŋ'geiʃən]	n.	伸长率，延伸率
thermoplastic	['θə:məu'plæstik]	a.	热塑性的
thermoset	['θə:məuset]	a.	热固性的
addition polymer			加成聚合物，加聚物
condensation polymer			缩合聚合物，缩聚物
hypothetical	[haipəu'θetikəl]	a.	假定的，理想的，有前提的
repeating unit		n.	重复单元
stereoregular	[stiəriə'regjulə]	a.	有规立构的，立构规整性的
stereoregularity	[stiəriəregju'lɑ:riti]	n.	立构规整性[度]
atactic	[ə'tæktik]	a.	无规立构的
isotactic	[aisəu'tæktik]	a.	等规[全同]立构的
polypropylene	[pɔli'prəupili:n]	n.	聚丙烯
polybutene	[pɔli'bju:ti:n]	n.	聚丁烯

Phrases

subdivide...in [into]... 把…细分为…，把…非正规地（划）分为…
range (from)...to... 落在（从）…到…之间；分布在从…到…范围内
be situated on [at, in]... 处于，位于，坐落于
exception to... …的例外（情况）
straighten out 拉直，打开
attraction for... 对…的引力
come under... 受…的影响［支配］

(be) subjected to... 经受…，受到…
in the order of... 大约…
from the mechanistic point of view 从反应机理的观点
be derived from... 由…产［派］生而来，来自于…
in a regular fashion 规则［整］地
bear in mind 牢记，记住
with the advent of... 随着…的出现
in relation to... 关于…，与…有关

Notes

① "As is found in all phases of polymer chemistry, there are many exceptions to this categorization." 句首为"as"引导的定语从句,说明整个主句,"as"在从句中作主语。介词短语"in all phases of polymer chemistry"在从句中作地点状语,其中的"phase"可作"方面、侧面、部分、阶段"解,可译作"在高分子化学的各个部分"。本句译文:"正如在高分子化学的各个部分都可以看到的那样,对这种分类方法有很多例外情况。"

② "When the molecular chains are 'straightened out' or stretched by a process of extension, they do not have sufficient attraction for each other to maintain the oriented state and will retract once the force is released." 在句首为以"when"引导的时间状语从句,其中"are 'straighten out'"和"are stretched"意义都是"被拉直",由于前者是从日常生活中借用来的故加引号。在主句中有两个谓语,即"do not have"及"will retract"。句中不定式短语"to maintain…"是宾语"attraction"的后置定语,句末"once"引导的从句为第二个谓语的时间状语。本句译文:"当通过一个拉伸过程将分子链拉直的时候,分子链彼此之间没有足够的相互吸引力来保持其定向状态,作用力一旦解除,将发生收缩。"

③ "Therefore, a potential fiber polymer will not become a fiber unless subjected to a 'drawing' process, i. e., a process resulting in a high degree of intermolecular orientation." 该句的主语为"a potential fiber polymer",意指"可以制成纤维的聚合物"。宾语"a fiber"后为一个条件状语从句"unless (it is) subjected to…",从句中省略了"it is"。"i. e."后为一个插入语,其中分词短语"resulting in…"为"a process"的后置定语。本句译文:"因此,可以制成纤维的聚合物将不成其为纤维,除非经受一个抽丝拉伸过程,即一个可以形成分子间高度取向的过程。"

④ "It must be borne in mind that, with the advent of Ziegler-Natta mechanisms and new techniques to improve and extend crystallinity, and the closeness of packing of chains, many older data given should be critically considered in relation to the stereoregular and crystalline structure." "It must be borne in mind"为主句,其中"it"为主句的形式主语,充当先行代词,代表"that"引导的从句,这个从句为主句的逻辑主语。从句中介词短语"with the advent of…"为从句的状语,这个短语中有两个介词宾语,即"mechanisms"和"techniques"。不定式短语"to improve and extend…"为"techniques"的后置定语,意指"提高…",这个不定式短语有两个宾语,即"crystallinity"和"closeness of packing"。从句的主语为"older data",意指"旧的资料",过去分词"given"为"data"的后置定语。在句子最后的介词短语"in relation to…"也为"data"的定语,为了使句子匀称,将其放在句子最后,此即"分割现象"。从句的谓语为"should be critically considered",可译作"应当批判地接受"。本句译文:"必须牢牢记住,随着 Ziegler-Natta 机理的出现,以及随着提高结晶度和提高链的堆砌密度的新方法的出现,对许多过去已得到的关于空间结构和晶体结构旧的资料,应当批判地接受。"

Exercises

1. *Translate the following passage into Chinese*

   In general, head-to-tail addition is considered to be the predominant mode of propagation in all polymerizations. However, when the substituents on the monomer are small (and do not offer appreciable steric hindrance to the approaching radical) or do not have a large resonance stabilizing effect, as in the case of fluorine atoms, sizable amounts of head-to-head propagation may occur. The effect of increasing polymerization temperature is to increase the amount of head-to-head placement. Increased temperature leads to less selective (more random) propagation but the effect is not large. Thus, the head-to-head content in poly (vinyl acetate) only increases from 1.30 to 1.98 percent when the polymerization temperature in increased from 30℃ to 90℃.

2. *Write out an abstract in English for the text in this unit*

3. *Put the following words into Chinese*

   entanglement   irregularity   sodium isopropylate   permeability   crystallite   stoichiometric balance fractionation   light scattering   matrix   diffraction

4. *Put the following words into English*

形态 酯化 异氰酸酯 杂质 二元胺 转化率 多分散性 力学性能 构象 红外光谱法

Reading Materials

Designing a Polymer Structure for Improved Properties

The three principles applied to give strength and resistance to polymers are (1) crystallization, (2) cross-linking, and (3) increasing inherent stiffness of polymer molecules. Combinations of any two or all of the three strengthening principles have proved effective in achieving various properties with polymers. For polymers composed of inherently flexible chains, crystallization and cross-linking are the only the available means to enhance polymer properties.

The third, and relatively new, strengthening principle is to increase chain stiffness. One possible way of stiffening a polymeric chain is to hang bulky side groups on the chain to restrict chain bending. For example, in polystyrene, benzene rings are attached to the carbon backbone of the chain; this causes stiffening of the molecule sufficient to make polystyrene a hard and rigid plastic with a softening temperature higher than that of polyethylene, even though the polymer is neither cross-linked nor crystalline. The method is advantageous because the absence of crystallinity makes the material completely transparent, and the absence of cross-linking makes it readily moldable. A similar example is poly (methyl methacrylate).

However, the disadvantage of attaching bulky side groups is that the material dissolves in solvents fairly easily and undergoes swelling, since the bulky side groups allow ready penetration by solvents and swelling agents. This problem can be eliminated by stiffening the backbone of the chain itself. One way to do this is to introduce rigid ring structures in the polymer chain. A classic example of such a polymer is cellulose, which is the structural frame-work of wood and the most abundant organic material in the world. Its chain molecule consisting of a string of ring-shaped condensed glucose molecules has an intrinsically stiff backbone. Cellulose therefore has a high tensile strength. In poly (ethylene terephthalate) fiber the chains are only moderately stiff, but the combination of chain stiffness and crystallization suffices to give the fiber high strength and a high melting point (265°C). The newer plastic polycarbonate containing benzene rings in the backbone of the polymer chain is so tough that it can withstand the blows of a hammer. Ladder polymers based on aromatic chains consist of double-stranded chains made up of benzene-type rings. These hard polymers are completely unmeltable and insoluble.

The combination of all three principles has led to the development of new and interesting products with enhanced properties. Composites based on epoxy- and urethane-type polymers may be cited as an example. thus, the stiff polymeric chains of epoxy and urethane types are cross-linked by curing reactions, and fillers are added to produce the equivalent of crystallization.

It is found that cold flow can be prevented by cross-links between the individual polymer chains. The structure of polymer chains present in the cross-linked polymers is similar to the wire structure in a bedspring, and chain mobility, which permits one chain to slip by another (which is responsible for cold flow) is prevented. Natural rubber, for example, is a sticky product with no cross-linking, and its polymer chains undergo unrestricted slippage; the product has limited use. However, as we have seen, when natural rubber is heated with sulfur, cross-linking takes place. Cross-linking by sulfur at about 5% of the possible sites gives rubber enough mechanical stability to be used in automobile tires but still enables it to retain flexibility. Introducing more sulfur introduces more cross-links and makes rubber inflexible and hard.

A high degree of cross-linking gives rise to three-dimensional or space network polymers in which all polymer chains are linked to form one giant molecule. Thus, instead of being composed of discrete molecules, a piece of highly cross-linked polymer constitutes, essentially, just one molecule. At high degrees of cross-linking, polymers acquire rigidity, dimensional stability, and resistance to heat and chemicals. Because of their network structure such polymers cannot be dissolved in solvents and cannot be softened by heat; strong heating only causes decomposition. Polymers or resins which are transformed into a cross-linked product, and thus take on a "set" on heating, are said to be of thermosetting type. Quite commonly, these materials are prepared, by intent, in only partially polymerized states (prepolymers), so that they may be deformed in the heated mold and then hardened by curing (cross-linking).

The most important thermosetting resins in current commercial applications are phenolic resins, amino-resins, epoxy resins, unsaturated polyester resins, urethane foams, and the alkyds. The conversion of an uncross-linked thermosetting resin into a cross-linked network is called curing. For curing, the resin is mixed with an appropriate hardener and heated. However, with some thermosetting systems (e.g., epoxies and polyesters), the cross-linked or network structure is formed even with little or no application of heat.

Aging of polymers is often accompanied by cross-linking due to the effect of the surroundings. Such cross-linking is undesirable because it greatly reduces the elasticity of the polymer, making it more brittle and hard. The well-known phenomenon of aging of polyethylene with loss of flexibility is due to cross-linking by oxygen under the catalytic action of sunlight. Cheap rubber undergoes a similar loss of flexibility with time due to oxidative cross-linking. This action may be discouraged by adding to the polymer an antioxidant, such as a phenolic compound, and an opaque filler, such as carbon black, to prevent entry of light.

——Chanda M, Roy S K. Plastics Technology Handbook. New York: Marcel Dekker, 1986. 33

Words and Expressions

stiffness	['stifnis]	n.	刚度[性], 硬度
strengthen	['streŋθən]	v.	加强, 增强, 强化
bulky	['bʌlki]	a.	体积庞大的
transparent	[træns'pɛərənt]	a.	透明的
rigid	['ridʒid]	a.	僵硬的
moldable	['məuldeibl]	a.	可模塑[制]的
structural framework			结构骨架
glucose	['glu:kəus]	n.	葡萄糖
intrinsically	[in'trinsikəli]	adv.	固有的, 本征[质]的
suffice	[sə'fais]	v.	足以, 足够
ladder polymer			梯形聚合物
epoxy	[e'pɔksi]	n.	环氧树脂
sticky	['stiki]	a.	黏性的, 胶黏的
slippage	['slipidʒ]	n.	滑动
prepolymer	[pri'pɔlimə]	n.	预聚物[体]
ag(e)ing	['eidʒiŋ]	n.	老化
opaque	['ɔpeik]	a.	不透明的

UNIT 10　Glass Transition Temperature

　　An ordinary rubber ball if cooled below −70℃ becomes so hard and brittle that it will break into pieces like a glass ball falling on a hard surface! Why does a rubber ball become like glass below −70℃? This is because there is a "temperature boundary" for almost all amorphous polymers (and many crystalline polymers) only above which the substance remains soft, flexible and rubbery and below which it becomes hard, brittle and glassy. This temperature, below which a polymer is hard and above which it is soft, is called the "glass transition temperature" T_g. The hard, brittle state is known as the glassy state and the soft, flexible state as the rubbery or viscoelastic state. On further heating, the polymer (if it is uncross-linked) becomes a highly viscous liquid and starts flowing; this state is termed viscofluid state, and the another transition takes place at its flow temperature T_f.

　　Now, let us consider a polymer, say, polyethylene. At room temperature, polyethylene is solid, exhibiting all characteristics of a low molecular weight substance. At high temperatures, however, the characteristic difference between high and low molecular weight substances can be seen. Depending on the temperature, the molecules of a low molecular weight substance either move apart as a whole or do not move at all, i.e., there is a definite temperature (melting point T_m) below which the molecules do not move and above which they do move. ①On the other hand, with polymers, if the temperature increases above T_g, localized units (chain segments) within the long chain molecule are first mobilized before the whole molecule starts moving. In some parts within the molecule, there is a considerable localized motion, but not in other parts of the same molecule. Thus, within the long chain of the polymer molecule, some segments have a certain freedom of movement, whereas others do not. The molecule as a whole does not move although some of its segments do.

　　In the case of polymers, there is indeed an intermediate state. If the temperature ranges between T_g and T_f, the localized mobility is activated but the overall mobility is not. The local segments, where mobility is already activated, correspond to the liquid state, while the molecule as a whole, where mobility is forbidden, is in the solid state. ② This state, which is really a combination of liquid and solid, is called the rubbery state. Under the influence of an applied stress, it exhibits properties of a viscous fluid as well as an elastic solid and undergoes what is called viscoelastic deformation. ③

　　The glass transition temperature T_g is an important parameter of a polymeric material. The T_g value of a polymer decides whether a polymer at the "use temperature" will behave like rubber or plastics. The T_g value along with the T_m value gives an indication of the temperature region at which a polymeric material trasforms from a rigid solid to a soft viscous state. ④ This helps in choosing the right processing temperature, i.e., the temperature region in which the material can be converted into finished products through different processing techniques such as moulding, calendering, extrusion, etc.

——Gowariker V R, Viswanathan N V, Sreedhar J. Polymer science. New York: John Wiley &Sons, 1986

Words and Expressions

brittle	['britl]	a.	脆的，易碎的
glass transition temperature			玻璃化转变温度
boundary	['baundri]	n.	界限，范围
flexible	['fleksəbl]	a.	柔软的
rubbery	['rʌbəri]	a.	橡胶状的
glassy	['glɑ:si]	a.	玻璃状的
glassy state			玻璃态
viscoelastic state			黏弹态
mobility	[məu'biliti]	n.	流动性
mobilize	['məubilaiz]	v.	运动，流动
segment	['segmənt]	n.	链段
deformation	[,di:fɔ:'meiʃən]	n.	形变
dimensional stability			尺寸稳定性
viscofluid state			黏流态
uncross-linked			非交联的
moulding	['məuldiŋ]	n.	模塑成型
calendering	['kælindəriŋ]	n.	压延成型
extrusion	[eks'tru:ʒən]	n.	挤出成型

Phrases

break into　摔碎
be known as　叫作
take place　出现，发生
depending on　根据
move apart　分开，分离，移开
as a whole　作为一个整体，整体（上）
correspond to　相当于

what is called　所谓的
transforms from...to...　由…转变成…
along with...　连同…，和……一起
give an indication of　表明，表示
be converted into　转化为

Notes

① Depending on the temperature, the molecules of a low molecular weight substance either move apart as a whole or do not move at all, i. e., there is a definite temperature (melting point T_m) below which the molecules do not move and above which they do move. 在此句中"depending on"作"根据"解。主句中"either…or…"作"要么…要么…"解；"move apart as a whole"应译成"整个分子移开"。"below which"和"above which"引导的从句均为"temperature"的定语从句。在该句末"they do move"中的"do"是强调动词"move"的。全句译文："根据温度，低分子量的物质要么整个分子移开，要么根本不运动，有一个特定的温度（熔点 T_m）在低于这一温度时分子不运动，而在高于这一温度时分子则在运动。"

② The local segments, where mobility is already activated, correspond to the liquid state, while the molecule as a whole, where mobility is forbidden, is in solid state. 在此句中"the local segments"为主语，意指"（在一条分子链上的）某些局部链段"。句中两个"where"引导的从句均为非限定性从句。"while"引导的从句为并列从句。"forbidden"本来作"被禁止"解，此处意指"没有被激活"。"is in"意为"处于"。全句译文："在一条分子链上某些流动性已经被激活的局部链段相应着流体状态，而没有被完全激活的整体分子则处于固体状态。"

③ Under the influence of an applied stress, it exhibits properties of a viscous fluid as well as an elastic solid and undergoes what is called viscoelastic deformation. 在此句中"it"代表"polymeric materials"。"undergo"作"进行"解，"what is called"为一个插入句，"undergoes what is called viscoelastic deformation"可译为"进行所谓的黏弹形变"。全句译文："在外加应力的作用下，聚合材料既显示出黏性液体的性质，也显示出弹性固体的性质，并发生所谓的黏弹形变。"

④ "The T_g value along with the T_m value gives an indication of the temperature region at which a polymeric material transforms from a rigid solid to a soft viscous state." 在此句中 "along with" 为 "和…一起"。"gives an indication of" 为 "表明"。"trasformsfrom...to..." 为 "由…转变成…"。全句译文: "T_g 值和 T_m 值表明了一个温度范围,在这个范围内聚合材料会由刚性固体转变为柔软的黏流态。"

Exercises

1. *Translate the following into Chinese*

If molecular geometry permits formation of a definite molecular orientation leading to a long-range three-dimentional order, the polymer has a greater crystallizability. That is why polymers with a symmetrical or stereo-regular structure are crystalline. The high crystalline polymers possessing a regular chain geometry show a high glass transition temperature because of their lower chain mobilities. Polymers with irregular chain backbone or randomly placed side groups are non-crystallizable. Segmental and chain mobilities are easier in non-crystallizable or amorphous polymers than in crystalline polymers, which leads the former to show a lower T_g.

2. *Give a definition for each following word*
    (1) glass transition temperature T_g
    (2) melting point T_m
    (3) flow temperature T_f
    (4) viscoelastic state
    (5) viscofluid state
    (6) glassy state

3. *Put the following words into Chinese*
    molecular motion   flexibility   orientation   backbone   single bond   bulky side group   intermolecular force   repeating unit   cohesive force   linkage

Reading Materials

Factors Influencing Glass Transition Temperature

The presence or absence of segmental and molecular motions decides whether a polymer is in a solid, rubbery or molten state. The nature and the magnitude of these motions depend on the size and geometry of the polymeric chain, flexibility of the chain segments and the type of molecular aggregates formed.

If molecular geometry permits formation of a definite molecular orientation leading to a long-range three-dimentional order, the polymer has a greater crystallizability. That is why polymers with a symmetrical or stereo-regular structure are crystalline. The high crystalline polymers possessing a regular chain geometry show a high glass transition temperature because of their lower chain mobilities. Polymers with irregular chain backbone or randomly placed side groups are non-crystallizable. Segmental and chain mobilities are easier in non-crystallizable or amorphous polymers than in crystalline polymers, which leads the former to show a lower T_g.

The glass transition temperature of a polymer is closely related to the flexibility of the chain. The segmental mobility of a polymer is determined by the degree of freedom with which different segments along the chain backbone can rotate around the covalent bonds. Linear polymer chains made of C—C, C—O or C—N single bonds have a high degree of freedom for rotation. The presence of aromatic or cyclic structure in the chain backbone or of bulky side groups on the backbone C atoms hinders the freedom for rotation. The higher the freedom to rotate is, the more flexible the chain segments are, and, hence, the higher their segmental mobility is. Similarly, the intermolecular forces determine the magnitude of the molecular aggregates. In the case of hydrocarbon polymers, only van der Waals' forces act on neighbouring chains and, hence, molecular aggregates are not that strong. Chain segments can slip past each other easily. Polymer chains containing polar groups, on the other hand, are held together more strongly by neighbouring dipole as well as intermolecular hydrogen bonding, and are unable to move that easily. Let us give some examples as follows:

1. Amorphous polyethylene, which is a hydrocarbon polymer and made up of repeating units of —CH_2—CH_2—, has a T_g of −125℃. In this case T_g is quite low because ① stong intermolecular cohesive forces are absent and ② the substituent group on C atoms is only hydrogen, which is not bulky at all.

2. Nylon 6, a polyamide, has a high glass transition temperature (50℃), primarily because of the presence of a large number of polar groups in the molecules, leading to strong intermolecular hydrogen bounding. It should be noted that intermolecular cohesive forces restrict segmental motion below 50℃ in the case of nylon 6.

3. The effect of the side group can be appreciated by comparing the T_g of polystyrene ($T_g=100℃$) with that of polymethylstyrene ($T_g=155℃$) and the T_g of polymethylacrylate ($T_g=8℃$) with that of polymethylmethacrylate ($T_g=105℃$). The presense of —CH_3 groups in poly-α-methylstyrene and polymethylmethacrylate comes in the way of free rotation around C—C bond of the chain backbone and, hence, hinders the chain mobility, resulting in an increase of 55℃ and 97℃ in their T_g values over polystyrene and polymethylacrylate, respectively.

4. A different trend in the effect of substituent side chains on T_g is seen if we examine the T_g values of methyl, ethyl and butyl polyacrylates. In this series of polyacrylates, the T_g values get reduced as the length of the side chain increases (i.e., 8℃, −22℃ and −54℃) for methyl, ethyl and butyl, respectively. Here also, the presence of the side chain, no doubt, hinders the free rotation of the C—C bond of the main chain, but this effect is almost the same in all the three polyacrylates. The side chains, nevertheless, due to the presence of C—C and C—O linkages, are themselves flexible. As the side chain length increases from methyl to butyl, their freedom of flexibility increases and they also tend to push the neighbouring main chains further apart, increasing thereby the free volume and, hence, the chain mobility.

5. A T_g value of 150℃ for polyvinyl carbazole is understandably high, as it contains bulky side groups which are directly linked on the chain backbone C atoms, as shown below:

6. In the case of polyethylene terephthalate, the presence of the aromatic ring in the chain backbone increases the inflexibility of the chain and the T_g value is high (69℃).

$$\sim O-\underset{H}{\overset{H}{C}}-\underset{H}{\overset{H}{C}}-O-\underset{O}{\overset{\|}{C}}-\!\!\!\left\langle\!\!\bigcirc\!\!\right\rangle\!\!-\underset{O}{\overset{\|}{C}}-O-\underset{H}{\overset{H}{C}}-\underset{H}{\overset{H}{C}}-O\sim$$

7. The higher glass transition temperature for the derivatives of cellulose, such as cellulose nitrate, is attributable to the rigid ring structure in the macromolecular chain.

8. The phenomenon of ultimate chain stiffening devoid of any segmental rotation may be seen in "ladder polymers" (the name derived from their structural resemlance to a ladder):

The ladder polymer has been found to be very stable thermally. Even upon direct exposure to a flame, the material shows a remarkable stability.

———Gowariker V R, Viswanathan N V, Sreedhar J. Polymer science. New York: John Wiley & Sons, 1986

Words and Expressions

aggregate	['ægrigeit]	n.	聚集体
geometry	[dʒi'ɔmitri]	n.	几何形状
substituent	[sʌb'stitʃuənt]	n.	取代基
hinder	['hində]	v.	阻碍
cohesive force			内聚力
restrict	[ri'strikt]	v.	限制,约束
appreciate	[ə'pri:ʃieit]	v.	了解,理解
linkage	['liŋkidʒ]	n.	键,键合
polyvinyl carbazole			聚乙烯咔唑
polyethylene terephthalate			聚对苯二甲酸乙二醇酯
cellulose nitrate			硝化纤维素
be attributable to			可归因于
stiffening	['stifniŋ]	n.	硬化,僵硬
devoid	[di'vɔid]	a.	全无的,缺乏的
devoid of			没有
ladder polymer			梯形聚合物
exposure	[iks'pəuʒə]	n.	暴露

UNIT 11 Functional Polymers

Functional polymers are macromolecules to which chemically functional groups are attached; they have the potential advantages of small molecules with the same functional groups. Their usefulness is related both to the functional groups and to the nature of the polymers whose characteristic properties depend mainly on the extraordinarily large size of the molecules. ①

The attachment of functional groups to a polymer is frequently the first step towards the preparation of a functional polymer for a specific use. ② However, the proper choice of the polymer is an important factor for successful application. In addition to the synthetic aliphatic and aromatic polymers, a wide range of natural polymers have also been functionalized and used as reactive materials. Inorganic polymers have also been modified with reactive functional groups and used in processes requiring severe service conditions. In principle, the active groups may be part of the polymer backbone or linked to a side chain as a pendant group either directly or via a spacer group. A required active functional group can be introduced onto a polymeric support chain (1) by incorporation during the synthesis of the support itself through homopolymerization or copolymerization of monomers containing the desired functional groups, (2) by chemical modification of a non-functionalized performed support matrix and (3) by a combination of (1) and (2). ③ Each of the two approaches has its own advantages and disadvantages, and one approach may be preferred for the preparation of a particular functional polymer when the other would be totally impractical. The choice between the two ways to the synthesis of functionalized polymers depends mainly on the required chemical and physical properties of the support for a specific application. Usually the requirements of the individual system must be thoroughly examined in order to take full advantage of each of the preparative techniques.

Rapid progress in the utilization of functionalized polymeric materials has been noted in the recent past. Interest in the field is being enhanced due to the possibility of creating systems that combine the unique properties of conventional active moieties and those of high molecular weight polymers. ④ The successful utilization of these polymers are based on the physical form, solvation behavior, porosity, chemical reactivity and stability of the polymers. The various types of functionalized polymers cover a broad range of chemical applications, including the polymeric reactants, catalysts, carriers, surfactants, stabilizers, ion-exchange resins, etc. In a variety of biological and biomedical fields, such as the pharmaceutical, agriculture, food industry and the like, they have become indispensable materials, especially in controlled release formulation of drugs and agrochemicals. Besides, these polymers are extensively used as the antioxidants, flame retardants, corrosion inhibitors, flocculating agents, antistatic agents and the other technological applications. In addition, the functional polymers possess broad application prospects in the high technology area as conductive materials, photosensitizers, nuclear track detectors, liquid crystals, the working substances for storage and conversion of solar energy, etc.

——Akelah A, Moet A. Functionalized Polymers and Their Applications. London: Chapman and Hall, 1990. 3

Words and Expressions

functional polymer			功能聚合物
aliphatic	[ˈælifætik]	a.	脂肪(族)的
inorganic polymer			无机聚合物
modify	[ˈmɔdifai]	v.	改性
backbone	[ˈbækbəun]	n.	主链，骨干
pendant group			侧基
spacer group			隔离基团
functionalized polymer			功能聚合物
porosity	[pɔːˈrɔsiti]	n.	多孔性，孔隙率
stability	[stəˈbiliti]	n.	稳定性
reactivity	[riækˈtiviti]	n.	反应性，活性
reactant	[riˈæktənt]	n.	反应物，试剂
carrier	[ˈkæriə]	n.	载体
surfactant	[səːˈfæktənt]	n.	表面活性剂
stabilizer	[ˈsteibilaizə]	n.	稳定剂
ion exchange resin			离子交换树脂
biological	[baiəˈlɔdʒikəl]	a.	生物(学)的
biomedical	[baiəˈmedikəl]	a.	生物医学的
pharmaceutical	[fɑːməˈsjuːtikəl]	n.	药品，药物
		a.	药物的，医药的
indispensable	[indisˈpensəbl]	a.	不可缺少的
controlled release			控制释放
drug	[drʌg]	n.	药品，药物
agrochemical	[ægrəuˈkemikl]	n.	农药，化肥
antioxidant	[æntiˈɔksidənt]	n.	抗氧剂
flame retardant			阻燃剂
corrosion inhibitor			缓蚀剂
flocculating agent			絮凝剂
antistatic agent			抗静电剂
conductive material			导电材料
photosensitizer	[fəutəuˈsensitaizə]	n.	光敏剂
nuclear track detector			核径迹探测器
liquid crystal			液晶
conversion	[kənˈvəːʃən]	n.	转化
solar energy			太阳能

Phrases

attach to... 和…连接
attachment of... to... 把…连接到…上，把…和…连接
the first step toward... 朝…迈进的第一步

take full advantage of... 充分利用…
interest in... …的(重大)意义，对…的关注

Notes

① "Their usefulness is related both to the functional groups and to the nature of the polymers whose characteristic properties depend mainly on the extraordinarily large size of the molecules." 主句的主语 "Their usefulness" 意指"它们(功能聚合物)的实用性"。"whose"引导的从句为主句中"the polymers"的定语从句。本句译文："它们(功能聚合物)之所以具有使用价值不仅与所带的官能团有关，而且与由

巨大的分子尺寸所决定的聚合物的特性有关。"

② "The attachment of functional groups to a polymer is frequently the first step towards the preparation of a functional polymer for a specific use." 主语部分 "the attachment of functional groups to a polymer" 意指 "把官能团连接到聚合物上"。在表语部分 "toward" 引导的介词短语为表语 "the first step" 的后置定语；而介词短语 "for a specific use" 则为 "a functional polymer" 的后置定语。本句译文："把官能团连接到聚合物上常常是制备某种特殊用途的聚合物的最重要的一步。"

③ "A required active functional group can be introduced onto a polymeric support chain (1) by incorporation during the synthesis of the support itself through polymerization or copolymerization of monomers containing the desired functional groups, (2) by chemical modification of a non-functionalized performed support matrix and (3) by a combination of (1) and (2)." 宾语 "a polymeric support chain" 可译作 "聚合物主链"。句中（1）、（2）、（3）三个由 "by" 引导的介词短语并列作方式方法状语。句中 "during" 引导的介词短语为时间状语。"through" 引导的介词短语为方式状语。句中 "a non-functionalized preformed support matrix" 可译作 "预先制成的未功能化的主链聚合物"。全句译文："可以用如下三种方法将所需要的活性官能团引入到聚合物主链上：（1）在合成主链聚合物时通过带有所需官能团的单体的均聚或共聚，使聚合物带上官能团；（2）将预先制成的未功能化的主链聚合物进行化学改性；（3）将（1）和（2）两种方法结合起来。"

④ "Interest in the field is being enhanced due to the possibility of creating systems that combine the unique properties of conventional active moieties and those of high molecular weight polymers." 本句主语 "Interest in the field" 意指 "对（功能聚合物）这个领域的关注"。谓语 "is being enhanced" 为现在进行时被动态，表明 "正在增加"。其后 "due to" 引导的介词短语作句子的原因状语。"systems" 代表 "功能聚合物"。"that" 引导的从句为 "systems" 的定语，其中的 "active moieties" 意指 "活性官能团"，"those" 代表前面出现的复数名词 "properties"。本句译文："由于能够制造出兼有活性官能团特性和高分子量聚合物性能的功能聚合物，所以人们对（功能聚合物）这个领域的兴趣与日俱增。"

Exercises

1. *Please answer the following questions in English*
   (1) What is called kinetic chain length?
   (2) What are addition polymers?
   (3) What are vinyl monomers?
   (4) What are puntionalized polymers?
2. *Please mark "yes" or "no" in the brankets*
   (1) Polyethylene is a typical vinyl polymers. (   )
   (2) The term nylon has been accepted as a generic commerical name for polyamides. (   )
   (3) Acetal is a type of polymers which is made from polyfunctional monomers with hydroxyl and carboxyl group. (   )
   (4) Polymethylmethacrylate is a kind of softer materials than atactic polypropylene at room temperature. (   )
   (5) The initial elastic modulus of fiber is greater both than rubber and than plastics. (   )
3. *Put the following words into Chinese*
   fragment   hydroxy acid   positive ion   reactivity ratio   irregularity   viscosity-average molecular weight   thermodynamically   random coil   dislocation   vulcanization
4. *Put the following words into English*
   合成橡胶   不饱和单体   双键   二元酸   偶合终止   平衡反应   异丁烯和苯乙烯的共聚物   分子间力   微晶   取向

Reading Materials

Natural Polymers

Three major classifications of natural polymers have occupied most of the literature in this field: polysaccharides, proteins, and polynucleotides. There are, however, other natural polymers, some of which have been in commercial use for a long time, that do not fit into any of these categories, and it would be instructive here to describe some of these materials.

A number of natural resins have been used for varnishes and molding compounds, with *shellac* being the best known. Shellac is a resin secreted by the lac insect, found in southern parts of Asia. It consists of a very complex mixture of crosslinked polyesters derived from hydroxy acids, principally aleuritic acid (9,10,16-trihydroxyhexadecanoic acid).

$$HO\text{---}(CH_2)_?\text{---}CH(OH)\text{---}CH(OH)\text{---}(CH_2)_?\text{---}COOH$$

Gas chromatographic analysis of the products of chemical degradation of shellac has also shown the presence of several saturated and unsaturated long-chain aliphatic acids together with other hydroxy-substituted acids and nonaliphatic compounds.

Some resins such as *sandarac* and *amber* are derived from trees. The latter is a fossil resin secreted in prehistoric times by certain types of evergreen trees. *Humic acid* is another type of fossil resin obtainable from fossil fuels, but also widely distributed in soils. Its structure is extremely complex and appears to consist mainly of highly substituted phenolic rings also containing carboxyl functions, and characterized by a high degree of hydrogen bonding and complexation with such other materials as proteins and polysaccharides. Partly because they are excellent scavengers of metals and because they appear to play a key role in soil drainage and movement of water, they have attracted the attention of environmental chemists in recent years. They already find some industrial use as pigment extenders and emulsifiers and for removal of boiler scale.

Rosin, which is obtained along with turpentine from pine trees as a component of the resinous discharge, is used in varnishes. Rosin is not polymeric, however, but consists of a mixture of monobasic substituted phenanthrene carboxylic acids. It is sometimes esterified with glycerol or pentaerythritol to form "ester gums" that are useful ingredients for adhesives and lacquers. (It will be recalled that another type of nonpolymeric natural product, drying oils, are also used in polymer modification, notably with polyesters.)

Asphalt or *bitumens* are resinous materials widely used in highway construction as aggregate binders——also for waterproofing buildings and as binders for roofing and flooring compositions. They occur in natural deposits but are obtained mainly from the residue from petroleum distillation.

Wood consists almost entirely of two materials, the polysaccharide cellulose and *lignin*. Lignin is a relatively low-molecular-weight polymer having phenolic rings linked through three-carbon units. The structure varies according to source, but an approximation of a segment of softwood lignin illustrates its complexity. Large quantities of lignin are produced as a by-product in paper-pulp manufacture, but it is of little use other than as an extender or binder in the production of pressed board (for example, Masonite). A number of attempts have been made to modify lignin to make it more suitable for commercial use—

for example, its copolymerization with styrene—but without much success.

One of the most important of all natural polymers is rubber (polyisoprene). The chemistry of this material was discussed in some detail in earlier publications and hence will not be repeated here; however, some of the structural characteristics of the various forms of natural rubber are of interest.

The principal form of natural rubber, which consists of 97% *cis*-1,4-polyisoprene, is known as *hevea rubber*. It is obtained by tapping the bark of a tree that grows wild in south America and is cultivated in other parts of the world. It is also found in some small shrubs and other plants. It is obtained as a latex consisting of about 32% to 35% rubber and about 5% of other compounds including fatty acids, sugars, proteins, sterols, esters, and salts. The polymer is of very high molecular weight (average about a million) and amorphous, although it becomes randomly crystallized at lower temperatures.

The latex can be converted to foam rubber by mechanical aeration followed by vulcanization. Rubber gloves and balloons are usually made by coating latex on forms before vulcanization. Most latex is coagulated (for example, with acetic acid) and used in bulk form. Most hevea rubber (about 65%) is used in tire manufacture, but it is also found in a host of commercial products including footwear, seals, weather stripping, shock absorbers, electrical insulation, sports accessories, and so on. All these utilize rubber in vulcanized form. One of the few applications of unvulcanized rubber is in the form of crepe which, because of its excellent abrasion resistance, is used for shoe soles.

Another form of natural rubber is *gutta-percha*, also obtained in latex form from trees (for example, *Palaquium oblongifolium* and similar trees mainly indigenous to Southeast Asia). Gutta-percha has the *trans*-1,4-polyisoprene structure. It is much harder and less soluble than hevea rubber and exists in crystalline form. It is used widely for golf ball covers.

Other forms of rubber related in structure to gutta-percha are *balata* and *chicle*, obtained from trees in Mexico and South and Central America. Balata is also used for golf ball covers, and as an impregnant for textile belting. Chicle is used in chewing gum, adhesive bandages, and varnishes.

——Stevens M P. Polymer Chemistry. London: Addison-Wesley Publishing Company, 1975. 385

Words and Expressions

natural polymer			天然聚合物
polynucleotide	[pɔli'nju:kliətaid]	n.	多核苷酸
instructive	[in'strʌktiv]	a.	有（教）益的
varnish	['vɑ:niʃ]	n.	清漆
shellac	[ʃə'læk]	n.	虫胶，紫胶片
secrete	[si'kri:t]	v.	分泌
lac insect			紫胶虫
aleurtic acid			紫胶酮酸
9,10,16-trihydroxyhexadecanoic acid		n.	9,10,16-三羟基十六酸
sandarac	['sændəræk]	n.	桧树胶
amber	['æmbə]	n.	琥珀
fossil resin		n.	琥珀树脂
prehistoric	['pri:histərik]	a.	很久以前的
evergreen tree			常绿树
humic acid			腐殖酸
complexation	[kɔmplek'seiʃən]	n.	络合（作用）
scavenger	['skævindʒə]	n.	清洗剂
drainage	['dreinidʒ]	n.	（水土）流失
rosin	['rɔzin]	n.	松香

turpentine	['tə:pəntain]	n.	松节油，松脂
phenanthrene	[fi'nænθri:n]	n.	菲
roofing	['ru:fiŋ]	n.	屋顶[面]材料
pentaerythritol	[pentəi'riθritɔl]	n.	季戊四醇
asphalt	['æsfælt]	n.	沥青，柏油
paper-pulp		n.	纸浆
binder	['baində]	n.	胶[黏]合剂
pressed board			压制板
Masonite	['meisənait]	n.	梅斯奈物纤维板
polyisoprene	[pɔli'aisəupri:n]	n.	聚异戊二烯
hevea	['hi:viə]	n.	三叶胶树
bark	[bɑ:k]	n.	树皮
cultivate	['kʌltiveit]	v.	种植
shrub	[ʃrʌb]	n.	灌木
sterol	['stərɔl]	n.	甾醇
latex	['leiteks]	n.	胶乳，乳液
aeration	[ɛə'reiʃən]	n.	充气，鼓风
indigenous	[indidʒi:nəs]	a.	野生的，土生土长的
balata	['bælətə]	n.	巴拉塔树胶
chicle	['tʃikəl]	n.	糖胶树胶
chewing gum			口香糖，橡皮糖

UNIT 12　Preparations of Amino Resins in Laboratory

Amino resins are reaction products of amino derivatives with aldehydes under acidic or basic conditions. The most important representatives of this class are the urea-formaldehyde (UF) and melamine-formaldehyde resins.

Chemicals. Urea, formalin (37%), ethanol, 2N NaOH solution, 0.1N NaOH solution, 1N standard NaOH solution, 1N standard HCl solution, glacial acetic acid, furfuryl alcohol, triethanolamine, wood flour, calcium phosphate, ammonium chloride, 0.5N H_2SO_4 solution, Na_2SO_3, 1% ethanolic thymolphthalein indicator solution, melamine, glycerol and monomethylolurea.

Apparatus. Flasks and beakers, 500 ml three-neck resin kettle, heating mantle, mechanical stirrer, condenser, Dean-Stark trap, oven, universal indicator paper, test tubes, 250-ml volumetric flask, ice bath, 10 ml pipette, burette, oil bath, screwcap and jar.

Preparation of a UF resin under acidic conditions. In order to demonstrate the rapidity of the reaction of urea with formaldehyde under acidic conditions, mix 5 g of urea with 6 ml of formalin in a test tube, and shake the tube until the urea has dissolved. Adjust the pH of the solution to 4 by the addition of 4 drops of 0.5N H_2SO_4, and observe the time required for precipitation to occur. Remove part of the precipitate and compare its solubility in water with the sample of monomethylolurea.

Preparation of a urea-formaldehyde adhesive. 60 grams (1 mole) of urea and 137 g (1.7 mole) of formalin (37%) are charged into a 500 ml reaction kettle equipped with a mechanical stirrer and a reflux condenser. The pH of the mixture is adjusted with 2N NaOH solution to between 7 and 8 as determined by universal indicator paper, and the mixture is refluxed for 2 hr.① At each subsequent 0.5-hr interval until the water has been removed, determine the free-formaldehyde content of the mixture by the procedure indicated below.②

After the mixture has refluxed for 2 hrs, a Dean-Stark trap is introduced between the flask and the reflux condenser. About 40 ml of water is distilled into the trap and is discarded. The solution is acidified with 5 drops of glacial acetic acid; 44 g of furfuryl alcohol and 0.55 g of triethanolamine are introduced into the reaction mixture, and the solution is heated at 90℃ for 15 min.

The mixture is cooled to room temperature. A 15 g sample of the resin is removed and is mixed with a hardener composed of 1 g of wood flour, 0.05 g of calcium phosphate, and 0.2 g of ammonium chloride. The mixture is set aside to harden at room temperature.③ The remaining resin to which the hardener has not been added is placed in a screwcap jar and is submitted to the laboratory instructor.

Determination of the free formaldehyde content. Prepare 250 ml of a 1M Na_2SO_3 solution, and neutralize the solution so it produces faint blue color with thymolphthalein indicator solution. Add a weighed 2 to 3 g sample of resin into 100 ml of water in a 250 ml Erlenmeyer flask, and swirl the contents of the flask until they are dissolved completely. If the resin will not dissolve, ethanol may be added to aid the solution process. Cool the solution to 4℃ in an ice bath. Place 25 ml of the 1M Na_2SO_3 solution in a 100 ml beaker, and pipet 10.00 ml of a standardized 1N HCl solution into the beaker. Cool the solution to 4 ℃. Add 10 to 15 drops of thymolphthalein indicator solution to the flask containing the sample, and adjust the color of the solution to a faint blue with 0.1N NaOH. Immediately, transfer the acid-sulfite solution to the flask containing the sample, completing the transfer with cold water. ④ Titrate the solution to the thymolphthalein endpoint with standard 1N NaOH solution.

$$CH_2O + Na_2SO_3 + H_2O \longrightarrow CH_2OHSO_3^- Na^+ + NaOH$$

Determine the % free formaldehyde from the quantity of HCl required to neutralize the resin solution.

Preparation of a melamine-formaldehyde resin. To a 500 ml reaction kettle equipped with a mechanical stirrer and a condenser is added 63 g (0.5 mole) of melamine and 122 g (1.5 mole) of formalin (37%). The mixture is refluxed for 40 min. The % of free formaldehyde should be determined but at 10-min intervals. The procedure for the free-formaldehyde determination is given above.

After 20 min heating of the sample, a Dean-Stark trap is inserted between the flask and the condenser, and 10 ml of water is distilled off. The uncured sample is placed in a screwcap jar and is submitted along with the cured resin to the laboratory instructor.

——McCaffery E L. Laboratory Preparation for Macromolecular Chemistry. New York: McGraw-Hill Book Company, 1970. 160

Words and Expressions

amino resin			氨基树脂
derivative	[di'rivətiv]	n.	衍生物
aldehyde	['ældihaid]	n.	醛
acidic	æ'sidik]	a.	酸(性)的
basic	['beisik]	a.	碱(性)的
urea-formaldehyde resin			脲醛树脂
melamine-formaldehyde resin			三聚氰胺甲醛树脂,蜜胺树脂
urea	['juəriə]	n.	尿素
formalin	['fɔ:məlin]	n.	福尔马林(40%甲醛水溶液)
ethanol	['eθənɔl]	n.	乙醇
glacial acetic acid			冰醋酸
furfuryl alcohol			糠醇
triethanolamine	[traieθə'nɔləmin]	n.	三乙醇胺
wood flour			木粉
calcium phosphate			磷酸钙
ammonium chloride			氯化铵
thymolphthalein	['θaimɔlfθæ'li:n]	n.	百里酚酞
indicator	['indikeitə]	n.	指示剂
glycerol	['glisərɔl]	n.	甘油
flask	[flɑ:sk]	n.	烧瓶
beaker	[bi:kə]	n.	烧杯
kettle	['ketl]	n.	釜,锅
mantle	['mæntl]	n.	套,加热套
stirrer	['stə:rə]	n.	搅拌器
condenser	[kən'densə]	n.	冷凝器
Dean-Stark trap			迪安-斯达克塔分水器
neutralize	['nju:trəlaiz]	v.	使…中和
test tube			试管
bath	[bɑ:θ]	n.	浴
pipette	[pi'pet]	n.	移液管
shake	[ʃeik]	v.	摇动

precipitation	[prisipi'teiʃən]	n.	沉淀
monomethylolurea	[mɔnəmeθilɔl'juəriə]	n.	单羟甲基脲
adhesive	[əd'hi:siv]	n.	胶黏剂
reflux	['ri:flʌks]	n.	回流
Erlenmeyer flask			锥形瓶
distill	[dis'til]	v.	蒸馏
acidify	[æ'sidifai]	v.	酸化
hardener	['hɑ:dnə]	n.	固化剂,硬化剂
harden	['hɑ:dn]	v.	固化,硬化
cured resin			固化树脂
acid sulfite			酸式亚硫酸盐
endpoint	['endpɔint]	n.	终点
titrate	['taitreit]	v.	滴定

Phrases

compare... with...　把…和…进行比较　　　　submit to　交给,提交
be charged into...　加入,投入,装入　　　　transfer... to...　把…转移到…
(be) equipped with...　装有…　　　　　　　cause... to...　使…做…
set aside　放置　　　　　　　　　　　　　　distill off...　把…蒸出

Notes

① "The pH of the mixture is adjusted with 2N NaOH solution to between 7 and 8 as determined by universal indicator paper, and the mixture is refluxed for 2 hr."该句子由两个并列分句构成。在第一个分句中"as determined by universal indicator paper"作方式状语,译作"通过用广泛试纸测定"。本句译文:"通过用广泛试纸测定用 2N NaOH 溶液把混合物的 pH 调至 7~8,然后将混合物回流 2h。"

② "At each subsequent 1/2 hr interval until the water has been removed, determine the free-formaldehyde content of the mixture by the procedure indicated below."该句为一个祈使句。以"at"引导的介词短语为方式方法状语。以"until"引导的从句为主句的时间状语。句末由"by"引导的介词短语为方式方法状语,其中"indicated below"为"Procedure"的后置定语。本句译文:"其后每隔半小时用下面的方法测定一次混合物中的自由甲醛含量,直到水完全脱除为止。"

③ "The mixture is set aside to harden at room temperature."句中不定式短语"to harden"在句子中作目的状语。介词短语"at room temperature"在句子中作条件状语。本句译文:"将混合物进行室温固化。"

④ "Immediately, transfer the acid-sulfite solution to the flask containing the sample, completing the transfer with cold water."该句话为一个祈使句。句中有两个"transfer",第一个为动词,意指"转移",而第二个则为名词,意指"热传递",也可译作"冷却"。"containing"引导的分词短语为"flask"的后置定语。由"completing"引导的分词短语作句子的时间状语。本句译文:"用冷水进行冷却以后,将酸式亚硫酸盐溶液立即转移到盛有试样的烧瓶中。"

Exercises

1. Put the correct words into the blanks in the following passage

When the center _____ radical activity has been transferred _____ the polymer _____ another molecule, the reaction is termed a transfer step. However, if the species formed does not have sufficient reactivity _____ add _____ monomer or _____ undergo any propagation reaction, a "degradative transfer" occurs. The reaction scheme _____ a radical undergoing a transfer reaction _____ a species X leading _____ the formation of an active free radical X· is _____ follows.

2. *Translate the following passage into English*

　　聚乙烯是一种广泛应用的聚合物,它具有很好的韧性、耐水性、耐溶剂性及抗污性。有三种形式的聚乙烯,即低密度聚乙烯、中密度聚乙烯和高密度聚乙烯。聚乙烯大量地用来制造管子、薄膜、电缆和其他用品。

3. *Describe the following symbols into English*

　　$\overline{DP}$; $\overline{M_n}$; $\overline{M_w}$; $\overline{M_v}$; MWD; $(\overline{r^2})^{\frac{1}{2}}$; $(\overline{S^2})^{\frac{1}{2}}$; HCl; BF$_3$; PVC; T_g; 100℃; 1.24g/cm^3; 75%; m.p.; 10^9 达因/cm^2.

4. *Put the following words into Chinese and write out their chemical structure.*

　　polymethylmethacrylate; polybutylene; polyamide; polyoxymethylene; titanium tetrachloride; ethylene oxide; ethylenediamine; isoprene; epichlorohydrin; chlorinated polyether

Reading Materials

Experimental Preparation of Polyester

Chemicals. Dibasic acids, anhydrides and diols listed in Table 12.1, pyrogallol, decalin, methanolic potassium hydroxide, and acetone.

Apparatus. Reaction kettle and appendages, heating mantle and rheostat, mechanical stirrer, distillation trap, condenser, nitrogen cylinder, thermocouple, pyrometer and asbestos gloves.

Table 12.1　Physical properties of compounds used in formation of polyesters

compound	boiling point/℃	melting point/℃
ethylene glycol	198	−12.6
diethylene glycol	245	−10.5
triethylene glycol	285	—
tetraethylene glycol	—	—
polyethylene glycol	—	determined by the molecular wt.
adipic acid	—	152
fumaric acid	200(sublimation)	287
maleic acid	135(decomposition)	130
maleic anhydride	202	53
phthalic acid	200(decomposition)	230
phthalic anhydride	284(sublimation)	130
succinic acid	235(decomposition)	185
succinic anhydride	261	119

Preparation of polyesters. A 1-liter four-necked reaction kettle provided with a Glascol heating mantle is charged with 35 ml of decalin and with one mole of the prescribed dibasic acid, or, if an anhydride is to be employed, the charge should contain 1/2 mole of glycol in addition to the anhydride and decalin. The rheostat should be adjusted to permit maximum safe heating of the mantle (usually about 110 volts) and the charge should be heated while the rest of the apparatus is being assembled.

　　The top of the reaction kettle is added and secured. The kettle is fitted with a mechanical stirrer, a 25-ml graduated distillation trap topped by a condenser, and a nitrogen-inlet tube which will permit the reactants to be blanketed with a stream of inert gas throughout the high-temperature polymerization. A thermocouple wire contained in a glass sleeve is connected to a pyrometer and is positioned so that the wire will make contact with the reaction

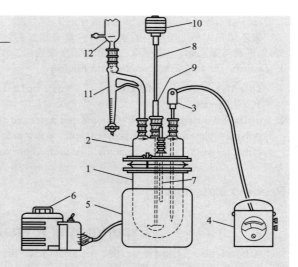

Fig. 12.1

Experimental device for preparing polyesters
1—reaction kettle bottom; 2—reaction kettle top; 3—thermocouple; 4—pyrometer; 5—heating mantle; 6—variable transformer; 7—nitrogen-inlet tube; 8—stainless steel stirrer; 9—stirrer bushing; 10—air motor; 11—graduated distillation trap; 12—condenser

mixture (see Fig. 12.1). The nitrogen to be used should be bubbled through a trap containing alkaline pyrogallol solution (50g of KOH, 100 ml of H_2O and 5g of pyrogallol) to remove oxygen and then through a tube of Drierite to remove water. The distillation trap should be filled with decalin to permit a reasonably constant weight to be maintained in the reaction flask as the reaction proceeds.

One mole of glycol (or the addition 1/2 mole of glycol if an anhydride is the coreactant) is preheated gently over a hotplate and when the temperature in the kettle has reached 150℃, the glycol is introduced into the reaction kettle through the arm containing the nitrogen-inlet tube. *Asbestos gloves should be worn for this and for all subsequent operations which require handling hot liquid.* Before replacing the nitrogen-inlet tube, add 0.17g (1 millimole) of *p*-toluenesulfonic acid to the reaction mixture. If the acidic comer is unsaturated, 0.1g of hydroquinone should also be added at this point.

The temperature of the reaction should be permitted to rise rapidly until reflux commences; this usually will occur before 180℃. Thereupon, the temperature should be maintained constant until one-fourth of the total water of reaction has been collected in the distillation trap. The water level should be recorded at 1-min intervals during this period. If the water-decalin interface is indistinct, a few drops of alkaline bromothymol blue indicator solution may be added to the trap; the indicator will be concentrated at the interface.

Once the first quarter of the water of reaction has been collected, the temperature of the reaction is permitted to rise about ten degree where the temperature is again maintained. The quantity of evolved water should be recorded at 1-min intervals both when the temperature is being increased and when it is being maintained. Be certain to record the temperature at which each reading is made. If water must be removed from the trap, replace the water with an equivalent quantity of decalin in order to maintain the system at constant volume. This procedure will help later to simplify your calculations.

After about half of the total water has been collected, stop the stirrer. Insert a glass tube into the kettle; using a suction device withdraw about 2 g of hot resin and deposit the sample into a small, weighed flask. *Be careful that the hot resin does not contact the skin.*

Replace the nitrogen-inlet tube, and permit the temperature of the reaction to rise rapidly an additional ten degree. The reaction is conducted isothermally at the new temperature until about three-quarters of the water has been evolved.

Meanwhile, the exact weight of withdrawn resin sample is determined. The resin is diluted with 10 ml of acetone, and after the sample has been dissolved, it is titrated with 0.8N methanoic KOH solution to a phenolphthalein endpoint. The extent of reaction as determined by titration should be compared with that calculated from the water of reaction.

Before the temperature is increased again, the stirrer is stopped and another sample of resin is withdrawn, weighed, and titrated. The stirrer is restarted, and the temperature of the reaction mixture quickly is raised another 10 degree where it is maintained until the water-evolution rate perceptibly is diminished. The reaction is completed isothermally at the highest practical temperature. The titration data are more meaningful than the azeotrope data in the later stages of the reaction. A final sample of resin should be titrated before the reaction is halted.

Some formulations cannot be permitted to react to their theoretical endpoint since side reactions cause crosslinking and its consequence-gelation. If the reaction mixture thickens markedly and it begins to ascend the stirrer shaft, additional glycol should be introduced quickly to induce ester-interchange reactions. This action will serve to inhibit the crosslinking reaction of the decomposing resin, and will facilitate the removal of the resin from the kettle. Measurements made beyond the gel point are meaningless; the partially gelled resin should be discarded.

When the reaction is concluded, the distillation trap and the nitrogen-inlet tube are removed. *Before proceeding further, be sure to put on the asbestos gloves*. The thermocouple and the stirrer may be wiped clean with a cotton swab once they have been removed. When the resin mixture has cooled below 200℃, it may be poured into a heavy-walled glass jar placed inside of a protective metal container. The conventional 8-oz screwcap laboratory jars are excellent for this purpose; they have good resistance to thermal shock.

The transfer of the resin into the jar is best accomplished by inverting the resin kettle while it is still secured in its mechanical mount and by permitting the kettle to drain in this position. The drained kettle also may be cleaned with cotton swabs.

The decalin-polyester mixture will separate upon cooling to room temperature. A characteristic resin sample will be obtained merely by decanting the supernatant decalin layer.

You should be able to calculate four different pseudo second-order rate constants(k') and four actual rate constants(k) from the data you have accumulated. Only those data secured isothermally should be included in your calculations. From the rate constants at the different temperatures you may determine the energy of activation (E) and the Arrhenius Factor (A) for your particular polymer. Compare the values for the parameters which you obtain with those obtained by the other groups for their respective polymers. Explain the significance of the Arrhenius parameters in relation to the structure of your polymer.

——McCaffery E L. Laboratory Preparation for Macromolecular Chemistry. New York: McGraw-Hill Book Company, 1970. 62

Words and Expressions

anhydride	[æn'haidraid]	n.	（酸）酐
pyrogallol	[paiərəu'gæləl]	n.	焦棓酚，焦性没食子酸，1,2,3-苯三酚
decalin	['dekəlin]	n.	十氢化萘，萘烷
methanolic potassium hydroxide			氢氧化钾甲醇溶液
ethylene glycol			乙二醇
diethylene glycol			一缩二乙二醇
triethylene glycol			二缩三乙二醇
tetraethylene glycol			三缩四乙二醇
polyethylene glycol			聚乙二醇
adipic acid			己二酸
fumaric acid			富马酸
sublimation	[sʌbli'meiʃən]	n.	升华
maleic acid			马来酸
phthalic acid			对苯二甲酸
succinic acid			丁二酸，琥珀酸
appendage	[ə'pendeidʒ]	n.	附件
rheostat	['riːəstæt]	n.	变阻箱［器］
cylinder	['silində]	n.	量筒，钢瓶
thermocouple	['θəːməukʌpl]	n.	热电偶
pyrometer	[paiə'rɔmitə]	n.	高温计
asbestos	[æz'bestəs]	n.; a.	石棉（的）
variable transformer			可调变压器
stainless steel			不锈钢
stream	[striːm]	n.	（气、液）流
sleeve	[sliːv]	n.	套管
bubble	['bʌbl]	v.	鼓泡
		n.	气泡
hotplate	['hɔtpleit]	n.	电炉
p-toluenesulfonic acid			对甲（基）苯磺酸
bromthymol blue			溴百里酚蓝，二溴百里酚磺酞
equivalent	[i'kwivələnt]	a.	等当量的
withdraw	[wið'drɔː]	v.	取样，取出
isothermally	[aisəu'θəːməli]	adv.	等温地
perceptibly	[pə'septibli]	adv.	显而易见地，显然
azeotrope	[ə'ziːətrəup]	n.	共沸混合物，恒沸物
ascend	[ə'send]	v.	升高
inhibit	[in'hibit]	v.	阻止，防止
discard	[dis'kɑːd]	v.	抛弃
drain	[drein]	v.	排出
pseudo	['psjuːdəu]	a.	假的

PART B

Polymerization Reaction Engineering

UNIT 13 Reactor Types

Reactors may be categorized in a variety of ways, each appropriate to a particular perspective. ① For example, Henglein (1969) chooses a breakdown based on the source of energy used to initiate the reaction (i.e., thermal, electrochemical, photochemical, nuclear). More common breakdowns are according to the types of vessels and flows that exist.

1. Batch reactors

The batch reactor (BR) is the almost universal choice in the chemist's laboratory where the most chemical processes originate. The reason is the simplicity and versatility of the batch reactor, whether it be a test tube, a three-neck flask, an autoclave, or a cell in a spectroscopic instrument. Regardless of the rate of the reaction, these are clearly low production rate devices. As scale up is desired, the most straightforward approach is to move to a larger batch reactor such as a large vat or tank.

Commercial batch reactors can be huge, 100000 gal or more. The cycle time, often a day or more, typically becomes longer as reactor volume increases in order to achieve a substantial production rate with an inherently slow reaction. ② Fabrication, shipping, or other factors place a limit the size of a batch reactor. For example, transportation capacity can limit the size of a batch reactor for which shop, as opposed to on-site, fabrication of the heat exchange surface is required. ③ This limits the production rates for which batch reactors may be economically utilized. Also, batch reactors must be filled, emptied, and cleaned. For fast reactions these unproductive operations consume far more time than the reaction itself and continuous processes can become more attractive.

2. Semibatch reactors (SBR)

Some reactions may yield a product in a different phase from the reaction mixture. Examples would be liberation of a gas from a liquid-phase reaction or the formation of a precipitate in a fluid-phase reaction. ④ To drive the reaction to completion, it may be desirable to continuously separate the raw product phase. A semi-batch operation may result as well from differing modes of feeding the individual reactants. ⑤ For reasons we will discuss later, it may be desirable to charge one reactant to the reactor at the outset and bleed a second reactant in continuously over time. Such reactors have both a batch and a flow character and, like batch reactors, are useful for slow reactions and low production rates.

3. Continuous stirred tank reactors (CSTR)

It is a small step from the batch reactor to the CSTR. The same stirred vessel may be used with only the addition of piping and storage tanks to provide for the continuous in- and outflow. Faster reactions can be accommodated and larger production rates can be achieved because of the uninterrupted operation. CSTRs are most often used for liquid-phase reactions, such as nitration and hydrolysis, and multiphase reactions involving liquid with gases and/or solids. Examples would be chlorination and hydrogenation.

4. CSTR in series

It was shown that considerable gains in production rate and economics can be achieved by passing the reacting mixture through a series of CSTRs. Again, we see how easy it is to achieve a gradual scaleup, say for a specialty chemical for which demand is increasing. ⑥ CSTRs in series are usually used for liquid-phase reactions.

5. Tubular reactors

As the production rate requirement increases, batteries of CSTRs become increasingly complex and tubular reactors become attractive. With the transition to tubular reactors, some versatility is lost and more process integration is required. Nevertheless, tubular reactors find extensive application in liquid-phase reactions, for example, polymerization, and are almost always the continuous reactor of choice for gas-phase reactions, for example, pyrolysis. Exceedingly high production rates can be achieved with tubular reactors either by increasing the diameter of the tube or more commonly by using a sufficient number of tubes in parallel.

6. Recycle reactors

Recycle reactor can be batch, CSTR, tubular, and so on in nature with the purpose of the recycle varying from one case to the next. Many large-scale commercial processes incorporate the recycle of one or more streams back to an earlier point in the process to conserve raw materials. This practice often results in the accumulation of impurities, which in turn requires separation. Usually it is not simply the reactor outlet stream that is recycled back to the reactor inlet, but it can be. ⑦ For example in a batch reactor the reacting mixture can be recycled, or pumped around, through a heat exchanger to provide thermal control.

Recycle reactors have also found valuable application in the laboratory and pilot plant because of their special characteristics. At one extreme, in which all of the product is recycled (no net flow), the reactor is the exact equivalent of the well-stirred batch reactor. ⑧ At the other extreme of no recycle, the reactor is simply the tubular variety. If there is some net flow but the recycle rate is high, the overall reactor performs like a CSTR. Yet the reaction tube itself behaves like differential tubular reactor. This versatility of the recycle reactor can be exploited to great advantage in research and development.

——Bisio A, Kabel R L. Scaleup of Chemical Processes. New York: John Wiley & Sons Inc., 1985. 255-257

Words and Expressions

categorize	['kætigəraiz]	v.	加以区别，分类
perspective	[pə:'spektiv]	n.	透视，观点

breakdown	['breikdaun]	n.	细目分类
batch reactor			间歇反应器
versatility	[və:sə'tiliti]	n.	多功能性，通用性，灵活性
three-neck flask			三颈瓶
autoclave	['ɔ:təkleiv]	n.	高压釜
cell	[sel]	n.	细胞，比色皿
spectroscopic	[spektrəs'kɔ:pik]	a.	分光镜的，光谱的
scale up			（反应器）放大
vat	[væt]	n.	大桶
fabrication	[fæbri'keiʃən]	n.	制造，装配
precipitate	[pri'sipitit]	n.	沉淀
raw product			粗产品
uninterrupted	[ʌnintə'rʌptid]	a.	连续的
nitration	['nait'reiʃən]	n.	硝化
hydrolysis	[hai'drɔ:lisis]	n.	水解
chlorination	[klɔ:ri'neiʃən]	n.	氯化
hydrogenation	[haidrədʒə'neiʃən]	n.	加氢
tubular reactor			管式反应器
integration	[inti'greiʃən]	n.	集成，综合
pyrolysis	[pai'rɔ:lisis]	n.	高温裂解
recycle reactor			循环反应器
incorporate	[in'kɔ:pə,reit]	v.	合并
conserve	[kən'sə:v]	v.; n.	保存，节省
differential tubular reactor			微分管式反应器
exploit	[iks'plɔit]	v.	开拓

Phrases

regardless of ＋ n.　不管…
at the outset　开始，开端
result from　因…而引起
say for ＋ n. / say that ＋ clause　例如…

CSTRs in series　多级串联连续流动搅拌反应器
batteries of　许多组（套）…
find extensive application in　起广泛作用

Notes

① "Reactors may be categorized in a variety of ways, each appropriate to a particular perspective." 译为："反应器可以用许多方法分类，各自适用于特定的目的。"

② "The cycle time, often a day or more, typically becomes longer as reactor volume increases in order to achieve a substantial production rate with an inherently slow reaction." 译为："对于慢化学反应，为了提高生产率必须增加反应器体积，而这往往导致反应器的循环周期变长，常常以天计算。"

③ "Fabrication, shipping, or other factors place a limit the size of a batch reactor. For example, transportation capacity can limit the size of a batch reactor for which shop, as opposed to on-site, fabrication of the heat exchange surface is required." 译为："制造、运输以及其他因素限制了反应器的规模，如热传递能力会限制间歇反应器的尺寸，热交换器必须在制造厂而不是在现场加工。"

④ "Examples would be liberation of a gas from a liquid-phase reaction or the formation of a precipitate in a fluid-phase reaction." 译为："例如液相反应中气体的释放或沉淀的生成。"

⑤ "A semibatch operation may result as well from differing modes of feeding the individual reactants." 译为："个别反应物的不同加入方式也导致半连续操作。"

⑥ "Again, we see how easy it is to achieve a gradual scaleup, say for a specialty chemical for which demand is increasing." 译为："另外，这种反应器容易逐步放大，例如某种化学品的需求逐步增加时常这样做。"

⑦ "Usually it is not simply the reactor outlet stream that is recycled back to the reactor inlet, but it can be."

译为："通常不是简单地将反应器的出料返回到入口，当然也可以这样做。"

⑧ "At one extreme, in which all of the product is recycled (no net flow), the reactor is the exact equivalent of the well-stirred batch reactor." 译为："一个极端是将所有的产物循环（没有净的流出），此时循环反应器严格等效于全混间歇反应器。"

Exercises

1. *Complete the following blanks according to the text*

Polymerizations involve highly complex reaction networks, producing highly complex molecular structures. Four types of reactors will be considered here: the _____ (BR), the semibatch (SBR), the plug flow (PFR), and the _____ _____ (CSTR). The concept of a batch reactor will be well known to readers. A _____ is one in which at least one reaction component is added over time. This may be a very small volume of material such as a _____ or _____, or the feed may comprise a large portion of the final volume of the reactor contents, as in the case of the semibatch addition of a comonomer.

The plug flow or _____ is, as the names imply, a tube through which the reaction fluid travels in _____ flow. The reactor may be a long, _____ tube, or a coil immersed in a heat transfer fluid. Since the reaction fluid travels in plug flow, there is no _____ mixing (no backmixing). Since the fluid must be in _____ flow to achieve the plug flow condition, radial mixing is assumed to be perfect. While these assumptions are never strictly correct, they off a realistic (approximate) description of a great many tubular reactors. With these assumptions, each element of _____ can be view as passing through the reactor without interacting with the elements before and after it.

2. *Put the following words into English*

间歇反应器　高压釜　三颈瓶　试管　连续搅拌反应器　平推流反应器　循环反应器　半连续反应器　光化学反应

3. *Put the following passage into Chinese*

Batch reactions are dictated for good quality in many polymerization schemes. Condensation polymerzations with exacting stoichiometric requirements are best done in batch weighings give tighter stoichiometry control than continuous flow metering. Reactions with a potential for crosslinking are also best done in batch since a flow reactor might eventually foul. Condensation polymers and living polymers such as anionidally polymerized vinyl addition polymers give the lowest polydispersity when done in batch. This limiting polydispersity is 2.0 for batch polycondensations carried to high conversion and 1.0 for batch anionic polymerizations which go to high molecular weights. Anionic polymerizations may foul flow reactors because molecular weights become arbitrarily high in the low-velocity regions near a wall.

Reading Materials

Polymerization Reactor Selection

The polymer "property" that all practicing engineers really strive to optimize is the bottom-line cost per unit production, subject to performance constraints. Meeting these performance constraints has been the subject of most of the concerned literature; however, cost can have important influence on the choice of polymerization reactor. Simplicity and energy efficiency is major goals along this route.

Considerations beyond cost and polymer product performance enter in as well. Viscosity of the reaction mixture is a primary consideration. Efficient mixing depends in part on power input, which in turn depends on viscosity (Oldshue, et al., 1982). An important caveat here is to take proper account of possible non-Newtonian rheological properties of the reaction mixture (Middleman, 1977). Flow properties largely determine agitator size, power, and design. Motionless mixers are also an important possibility to consider (Middleman, 1977). Conveying and pumping equipment must be designed around the flow properties of the reaction mixture. Extruders can be used as pumps and mixers. In many cases, the flow properties in the reactor have determined historically the choice of reactor for particular purpose. The rationale for these choices is found in the patent literature, which has been reviewed nicely up to the late 1970s by Gerrens (1982).

Separation processes that may have to be accomplished after the polymerization reaction can influence the choice of reactor. Two principal separation operations follow polymerization processes: elimination of monomers or diluents and separation of solid polymer. Devolatilization, to reduce residual monomer content to negligible levels, is an important example of the former. Extruders, tower, and wiped film reactors are all used in processes where devolatilization is a chief concern. There are also many common design concerns between devolatilization and reactor designs for driving certain step-growth polymerization to very high conversion by removal of condensation product (pressure reduction, surface area regulation). Effective separation of solid product has been one of the important innovations in certain reactors, such as the Union Carbide fluidized-bed reactor process for polyolefins. Loop reactors incorporating a settling portion are also useful for separating solid polymer from slurries.

Temperature and its control are always considerations. Heats of polymerization are typically high, as mentioned earlier, so that maintaining desired temperature is not always a simple matter. Temperature can become spatially nonuniform and globally out of control. The typical consistency of the reaction medium is again a factor. Normal good heat transfer design for large heat transfer surface, with coils or corrugated surfaces, is not effective in polymerization reactors since these surfaces create dead zones and accumulate material. This can produce nonuniform molecular weight distribution; foul the heat transfer equipment, exacerbating the heat removal difficulties; and make the reactor very difficult to clean. For these reasons, smooth heat transfer surfaces are usually preferable, with the best possible agitation to sweep the fluid clearly over the surface.

An example of these considerations can be seen in high-pressure ethylene polymerization. This reaction is done commercially in both tubular and stirred autoclave reactors. Tubular reactors have a high surface-to-volume ratio, which is good for heat transfer; they have, however, no mixing, which may produce segregation and inhomogeneity. Stirred-tank reactors have a comparatively low surface-to-volume ratio, although temperature can be manipulated somewhat using the feed temperature. Generally, these reactors have the better mixing.

The desired form (pellet, powder, bead, etc.) can influence the reactor design. For example, suspension polymerizers produce beads that may be directly useful in processing. On the other hand, round beads can be dangerous if spilled and may not have suitable bulk flow properties. Extruder reactors (Stuber and Tirrell, 1985) are able to produce pellets, sheets, or coatings quite easily and directly. Safety considerations always place inviolable constraints on polymerization process design. Venting to the atmosphere in a safe and environmentally sound manner must always be designed for. Clearly, the entire process, not exclusively the reaction, must be considered in any useful polymerization process optimization. The articles by Platzer (1970) and Gerrens (1982) give additional useful information and insight into the practical aspects of reactor selection.

——Tirrell M, Galvan R, Laurence R L. Polymerization Reactors in: Carberry J J, Varma A, ed. Chemical Reaction and Reactor Engineering. New York and Basel: Marcel Dekker, Inc., 1988. 772-773

Words and Expressions

strive	[straiv]	v.	努力
constraint	[kən'streit]	n.	约束
caveat	['keiviˌæt]	n.	告诫，警告
rheological	[ˌriːəˈlɔdʒikəl]	a.	流变的
agitator	['ædʒiteitə]	n.	搅拌器
extruder	[iks'truːdə]	n.	挤出机
rationale	[ˌræʃəˈnɑːl]	n.	说明，基本原理
diluent	['diljuənt]	n.	稀释剂
devolatilization	[diːvɔˌlætilaiˈzeiʃən]	n.	脱挥
polyolefin	[ˌpɔliˈəuləfin]	n.	聚烯烃
settling	['setliŋ]	n.	沉淀物
nonuniform	[ˌnɔnˈjuːnifɔːm]	a.	不均匀
corrugate	['kɔːrəgeit]	a.	波纹状的
foul	[faul]	v.	淤塞，弄脏，结垢
exacerbate	[eksˈæsəːbeit]	v.	恶化
ethylene	['eθiliːn]	n.	乙烯
inhomogeneity	[ˈinhɔməudʒeˈniːti]	n.	不均一(性)，多相(性)
pellet	['pelit]	n.	小球
polymerizer	['pɔliməraizə]	n.	聚合反应器
spill	[spil]	n.; v.	溢出
inviolable	[inˈvaiələbl]	a.	神圣的,不容忽视的
vent	[vent]	v.	排放
sound	[saund]	a.	可靠的,合理的
influence on			对…有影响
loop reactor			环管反应器
motionless mixer			静态混合器
out of control			失控
Union Carbide			联碳公司(简称 UCC)
fluidized-bed reactor			流化床反应器
take proper account of			正确考虑…
drive…to…			迫使
insight into			深刻了解

UNIT 14　Bulk Polymerization

Bulk polymerization traditionally has been defined as the formation of polymer from pure, undiluted monomers. Incidental amounts of solvents and small amounts of catalysts, promoters, and chain-transfer agents may also be present according to the classical definition. This definition, however, serves little practical purpose. It includes a wide variety of polymers and polymerization schemes that have little in common, particularly from the viewpoint of reactor design. The modern gas-phase process for polyethylene satisfies the classical definition, yet is a far cry from the methyl methacrylate and styrene polymerization which remain single-phase throughout the polymerization and are more typically thought of as being bulk. ①

A common feature of most bulk polymerization and other processes not traditionally classified as such is the need to process fluids of very high viscosity. The high viscosity results from the presence of dissolved polymer in a continuous liquid phase. Significant concentrations of a high molecular-weight polymer typically increase fluid viscosities by 10^4 or more compared to the unreacted monomers. This suggests classifying a polymerization as bulk whenever a substantial concentration of polymer occurs in the continuous phase. Although this definition encompasses a wide variety of polymerization mechanisms, it leads to unifying concepts in reactor design. The design engineer must confront the polymer in its most intractable form, i.e., as a high viscosity solution or polymer melt.

The revised definition makes no sharp distinction between bulk and solution polymerizations and thus reflects industrial practice. Several so-called bulk processes for polystyrene and ABS② use 5%～15% solvent as a processing aid and chain-transfer agent. Few successful processes have used the very large amounts of solvent needed to avoid high viscosities in the continuous phase, although this approach is sometimes used for laboratory preparations.

Bulk polymerizations often exhibit a second, discontinuous phase. They frequently exhibit high exothermicity, but this is more characteristic of the reaction mechanism than of bulk polymerization as such. Bulk polymerizations of the free-radical variety are most common, although several commercially important condensation processes satisfy the revised definition of a bulk polymerization.

In all bulk polymerizations, highly viscous polymer solutions and melts are handled. This fact tends to govern the process design and to a lesser extent, the process economics. ③ Suitably robust equipment has been developed for the various processing steps, including stirred-tank and tubular reactors, flash devolatilizers, extruder reactors, and extruder devolatilizers. Equipment costs are high based on working volume, but the volumetric efficiency of bulk polymerizations is also high. If a polymer can be made in bulk, manufacturing economics will most likely favor this approach. ④

It is tempting to suggest that polymer processes will gradually evolve toward bulk. ⑤ Recently, the suspension process for impact polystyrene has been supplanted by the bulk process, and the emulsion process for ABS may similarly be replaced. However, the modern gas-phase process for polyethylene appears to represent an opposite trend. It seems that polymerization technology tends to eliminate solvents and suspending fluids other than the monomers themselves. When the monomer is a solvent for the polymer, bulk processes as described in this article are chosen. When the monomer is not a solvent, suspension and slurry processes like those for polyethylene and polypropylene are employed. Hence, it is worthwhile avoiding a highly viscous continuous phase, but not at the price of introducing extraneous material. ⑥

——Nauman E B. Enlyclopedia of Polymer Science and Engineering. 2nd ed. Vol. 2. Editor-in-chilf Kroschwitz J I. New York: John Wiley & sons, 1985. 500-501

Words and Expressions

bulk polymerization			本体聚合
promoter	[prə'məutə]		促进剂
methylmethacrylate		n.	甲基丙烯酸甲酯
concentration	[ˌkɔnsen'treiʃən]	n.	浓度
solution polymerization			溶液聚合
substantial	[səb'stænʃəl]	a.	实质的,真实的
confront	[kən'frʌnt]	v.	面临
intractable	[in'træktəbl]	a.	难处理的
exothermicity	[ˌeksəu'θɜːmisiti]	n.	放热性
robust	[rəu'bʌst]	a.	强壮的,精力充沛的,坚固耐用的
devolatilizer	[di'vɔlætilaizə]	n.	脱挥器
tempt	[tempt]	v.	诱惑,感兴趣
impact polystyrene			抗冲聚苯乙烯
supplant	[sə'plɑːnt]	n.	代替

Phrases

so-called   所谓的
incidental / small amounts of   少量的
from the viewpoint of   从…观点
be a far cry from   与…截然不同

more compared to   与…比较有明显差别
make no sharp distinction between… and…   没有明显区别

Notes

① "The modern gas-phase process for polyethylene satisfies the classical definition, yet is a far cry from the methyl methacrylate and styrene polymerization which remain single-phase throughout the polymerization and are more typically thought of as being bulk." 译为:"根据传统的定义,现代的气相聚乙烯工艺属于本体聚合过程,但与通常认为的甲基丙烯酸甲酯和苯乙烯的典型本体聚合过程截然不同,后者在整个聚合反应过程中始终保持单相体系。"

② "ABS acrylonitrile-butadiene-styrene terpolymer." 译为:"丙烯腈-丁二烯-苯乙烯三元共聚物"。ABS 是一种工程塑料。

③ "This fact tends to govern the process design and to a lesser extent, the process economics." 译为:"(本体聚合过程的高黏现象)支配聚合过程的设计,使过程经济效益下降。"

④ "If a polymer can be made in bulk, manufacturing economics will most likely favor this approach." 译为:"如果某一聚合物能通过本体聚合过程制造,从经济学角度该过程是最适宜的。"

⑤ "It is tempting to suggest that polymer processes will gradually evolve toward bulk." 译为:"聚合物生产向本体聚合发展非常具有吸引力。"注意句型 It is tempting to suggest that…

⑥ "Hence, it is worthwhile avoiding a highly viscous continuous phase, but not at the price of introducing extraneous material." 译为:"(淤浆聚合)能避免高黏度连续相,又无其他物料导入的问题。"

Exercises

1. *Complete the following blanks according to the text*

The common characteristic of bulk _____ is the presence of high molecular-weight polymer in the _____ phase. Significant _____ of polymer lead to very high viscosities, which give rise to laminar-flow systems with mass- and heat-transfer limitations. The heat-transfer limitations are primarily attributable to existence of _____ flow and secondarily to lowered molecular diffusivities. Because of reduced diffusivities, the Trommsdorf effect is common in _____ polymerizations. The Combination of _____ and _____ leads to segregation effects in continuous-flow reactors, affecting molecular-weight and copolymer-composition distributions. Heat-transfer _____ are common in production-scale equipment used for free-radical bulk polymerization.

2. *Put the following words into English*

本体聚合　溶液聚合　表观黏度　流变　高抗冲聚苯乙烯　ABS　放热性　缩聚过程　聚乙烯
聚丙烯　聚烯烃　甲基丙烯酸甲酯

3. *Put the following passage into Chinese*

Most high-viscosity polymerizations require close-clearance impellers. These impellers have average shear rate constants around 30. Although high-viscosity impellers such as anchors or helical impellers operate at average of 10 r/min in high-viscosity installations, the shear stress upon the particles is very high. In fact, in most solution and bulk polymerization the shear rate is low but the shear stress is high, due to the high viscosity.

BOLTZMANN LIQUIDS: The apparent viscosity of these liquids varies as a function of the time constant shear stresses are applied. Depending on whether their apparent viscosity decreases or increases, they are known respectively as thixotropic or rheopectic liquids. To allow for hysteresis effects, rhelolgical measurements must be taken at an increasing and then decreasing velocity gradient.

Reading Materials

Polymerization Viscosities

Viscosity in itself is a nebulous term when describing the polymer or polymer solution in most polymerizations. Some polymerizations are carried out in water with small beads being formed and suspended in the water. The "viscosity" of such a system could actually mean the viscosity of the water, the viscosity of the slurry present with the beads in the water, the impeller viscosity, the process viscosity, the bulk viscosity, or the viscosity at the heat transfer surface.

In the bulk polymerization method, knowledge of viscosity is of vital importance. Bulk polymerizations typically operate between 100000 and 500000 cP (100 and 500Pa · s) bulk viscosity. An accurate determination of the bulk viscosity is extremely important in addition to the rheology associated with the particular polymer. Because bulk polymerizations are generally high viscosity in nature, the corresponding mixing Reynolds number is very low, normally less than 100. This is in the laminar region. Power is proportional to $N^2 D^3 \mu$ in the laminar range; so the actual horsepower which the mixer will draw is proportional to viscosity. Because of this, it is a requirement that viscosity vs. shear rate data be known. For example, assume two separate companies manufacturing bulk polystyrene have presented viscosity data to mixer vendors. Both companies have stated that the bulk viscosity of this material is 300000 cP (300Pa · s). Company A furnished only this information to the mixer suppliers. Company B furnished the bulk viscosity information in addition to the viscosity vs. shear rate data. Because the mixer manufacturer could determine the proper viscosity to load the impeller from the information that customer B furnished, the mixer recommended was a 25hp design (18.5 kW). Company A received a quote for a 50-hp mixer (37kW). Both mixers were for tanks of the same size and shape and operating at the same speed. Naturally the quotation for customer A will be at a higher price and a higher operating cost than for customer B. However, both mixers will accomplish the required results.

About 85% of all high-viscosity materials are pseudoplastic and viscoelastic in nature. Bulk polystyrene, polyesters, and polyelectrolytes are pseudoplastic in nature. Most materials have a slope of -0.6 to -0.2 when viscosity is plotted vs. shear rate. By reviewing these data and comparing the viscosity-vs.-shear rate information with the known shear rate constant of close-clearance impellers, the impeller viscosity can be determined. The shear rate constant for anchors and helical impellers is 30.

As indicated earlier, helical impellers and anchors are typically used in bulk polymerizations. However, neither of these two devices can operate effectively without viscous drag at the wall of the vessel. Without some drag the material in the tank will turn as one entire mass, and almost no mixing will occur. Therefore, in bulk polymerizations it is important to be sure that inlet pipes of low-viscosity material and reflux lines are directed toward a point at the liquid level one-half of the distance from the mixer shaft to the tank wall. This will allow incorporation of the low-viscosity material and prevent its migration to the tank walls where it could act as a lubricating layer, thereby reducing the agitation. It is also important to optimize the temperature differential ΔT between the bulk fluid and the heat transfer surface. Normally bulk viscosity applications only require tank jackets to obtain temperature control. A very high jacket temperature could reduce the viscosity of the material at the tank wall to a point where it acts as a lubricating boundary layer. Too cold a temperature at the tank wall could increase the viscosity dramatically to a point where the mixer is not designed to handle it. In this case a totally stagnant boundary layer at the wall could occur and product quality could be affected. Furthermore, damage could result to the mixer, drive motor, and vessel.

Viscosities for solution polymerizations are normally 25000 and 500000 cP (25 and 500Pa·s) bulk viscosity. The same problems exist with the term viscosity in this type of polymerization as in bulk polymerizations. An exact knowledge of the bulk viscosity and viscosity at the impeller are important. In lower viscosity materials, where open impellers are used, the importance of the viscosity determination is slightly reduced because the mixing Reynolds numbers are normally in the transition region where horsepower is not proportional to viscosity. Therefore, a minor change in the viscosity will have little effect on the horsepower drawn by the mixer.

Most solution polymerizations use tank jackets as heat transfer media; however, some solution polymerizations require additional surface area. Again, the ΔT optimization is important in solution polymerizations.

——Oldshue J Y. Fluid Mixing Technology. New York: McGraw-Hill Pub. Co., 1983. 462-464

Words and Expressions

nebulous	['nebjuləs]	a.	模糊的
rheology	[ri:'ɔlədʒi]	n.	流变学
laminar	['læminə]	a.	层流的
vendor	['vendɔ:]	n.	卖方
supplier	['sʌplaiə]	n.	供方
quote	[kwəut]	vt.	报价
viscoelastic	[,viskəui'læstik]	a.	黏弹性的
polyelectrolyte	[,pɔli'lektrəlait]	n.	高分子电解质
slope	[sləup]	n.	斜率
anchor	['æŋkə]	n.	锚式搅拌器
migration	[mai'greiʃən]	n.	移动
lubricate	['lu:brə,keit]	v.	润滑
jacket	['dʒækit]	n.	夹套
dramatically	[drə'mætikəli]	adv.	戏剧性地、显著地
stagnant	['stægnənt]	a.	停滞的
reflux	[ri:'flʌks]	n.	逆流,回流
viscosimeter	[,viskəu'simitə]	n.	黏度计
viscosity-vs-shear rate			黏度与剪切速率
shear rate constant			剪切速率常数
close-clearance impeller			近壁式搅拌桨
helical impeller			螺带式搅拌桨
inlet pipe			进料管线
boundary layer			边界层
open impeller			开式搅拌桨
transition region			过渡流域
turbulent region			湍流域
mixing viscosimeter			搅拌桨式黏度计

注: mixing Reynolds number

搅拌雷诺准数,$Re = \dfrac{\rho N d^2}{\mu}$ 其中ρ为密度,μ为黏度,N为搅拌转速,d为搅拌桨直径。

UNIT 15　General Description of VC Suspension Polymeri-zation Process

By far the majority of PVC is produced via the suspension route. ① In this process the vinyl chloride monomer is suspended, as liquid droplets, in a continuous water phase by a combination of vigorous agitation and the presence of a protective colloid (dispersant or suspending agent). A monomer soluble free radical initiator is used such that the polymerization takes place within the suspended droplets, via a mechanism that has been shown to be equivalent to that found in bulk polymerization.

Commercial plants are based on batch reactors, the size of which have increased progressively over the years. The original plants built during the 1940s usually consisted of 1000 gallon reactors. During the 1950s and 1960s this increased to 3000~5000 gallon and subsequently, during the early 1970s, 29000 gallon reactor systems were developed by Shin Etsu② and up to 44000 gallon (200m^3) by the German company Huls. At the present time few new plants being built consist of reactors of less than 15000 gallon capacity, having a batch size of about 25 tonnes of monomer. Smaller reactors were often glass lined to give a smooth finish that would resist the laydown of deposits on the walls. ③ Larger reactors are usually of polished stainless steel. ④

The polymerization of vinyl chloride is a very exothermic reaction and so the ability to remove heat is usually the limiting factor in attempting to minimise reaction times. As the size of reactors has increased the surface area to volume ratio has decreased thus aggravating the problem. Internal cooling coils are not usually used as they would attract deposits and be difficult to clean and thus have an adverse effect on product properties. The problem is often overcome by the use of chilled water or the fitting of a reflux condenser which, via the continued refluxing of vinyl chloride monomer, utilises its latent heat of vapourisation for cooling purposes. ⑤

A simple suspension polymerization formulation might consist of the following ingredients:

ingredient	typical level(parts by weight)
water	150~200
vinyl chloride	100
pH regulator	0.05
dispersant(s)	0.1
initiator(s)	0.06

Cold water is usually charged to the reactor first although it is sometimes preheated. The pH regulator is then added followed by the dispersant (s) in the form of a pre-prepared solution. The initiator (s) are sprinkled onto the surface of the aqueous phase immediately before sealing the reactor which is then evacuated to remove oxygen as this can increase polymerization time and affect product properties. When evacuation is complete the vinyl chloride is charged and heat-up of the reactor contents commenced. ⑥ Reaction temperature, which is the main controlling factor for product molecular weight, is commonly in the range 50~70℃ resulting in reactor pressures of 100~165 psi.

The trend is towards the operation of large closed reactors which are only opened for maintenance or, possibly, occasional cleaning operations. ⑦ In this case all ingredients are charged as solutions or dispersions and there would generally be no need for the evacuation step.

When the desired conversion has been reached, usually 75%~95%, the reaction can be chemically short-stopped if required and the bulk of the remaining monomer recovered. The product slurry is then stripped down to very low residual vinyl chloride levels by treatment at elevated temperatures, either in the reactor or similar vessel, or by contact with steam in a countercurrent multiplate stripping tower. The slurry is then dewatered by centrifuging and the resulting wet cake dried, commonly on a multi-stage flash drier, although a wide variety of different drier types is used by various manufacturers. After drying, the product is passed through some kind of scalping screen to remove extraneous large particles before bagging or loading to bulk road tankers.

——Butters G, ed. Particulate Nature of PVC. London: Applied Science Pub. Ltd., 1982. 4-6

Words and Expressions

vinyl chloride			氯乙烯，常缩写为 VC
colloid	['kɔlɔid]	n.	胶体
dispersant	[dis'pə:sənt]	n.	分散剂
line	[lain]	v.	衬里，贴面
deposit	[di'pɔzit]	n.; v.	堆积物，沉淀
minimise	[,minimaiz]	v.	最小化（美语为 minimize）
aggravate	['ægrəveit]	v.	加重，恶化
chilled water			冷冻水
latent	['leitənt]	a.	潜在的
formulation	[fɔ:mju'leiʃən]	n.	配方
ingredient	[in'gri:djənt]	n.	成分
sprinkle	['spriŋkl]	v.	喷洒
seal	[si:l]	v.	密封
evacuate	[i'vækjueit]	v.	撤出
commence	[kə'mens]	v.	开始，着手
slurry	['slə:ri]	n.	淤浆
countercurrent	[,kauntəkʌrənt]	n.	逆流
stripping tower			脱单塔
dewater	[di'wɔ:tər]	v.	脱水
centrifuge	['sentrə,fju:dʒ]	v.	离心
scalp	['skælp]	n.; v.	筛子，筛分
tanker	['tæŋkə]	n.	油轮，槽车
manufacturer	[mænju'fæktʃərə]	n.	生产者，生产商，生产厂家

Phrases

by far　非常
have an adverse effect on　对…不利

strip down　脱去

Notes

① "By far the majority of PVC is produced via the suspension route." 译为:"绝大多数聚氯乙烯是采用悬浮聚合工艺生产的。""By far"非常,起强调作用。

② Shin Etsu 日本信越公司,开发成功 125 m³ 的氯乙烯悬浮聚合反应器。中国齐鲁石化、上海氯碱从日本引进了该技术。

③ "Smaller reactors were often glass lined to give a smooth finish that would resist the laydown of deposits on the walls." 译为:"小型的(氯乙烯悬浮聚合)反应器内壁搪玻璃使壁面光滑,以防止聚合物粘壁。"

④ "Larger reactors are usually of polished stainless steel." 译为:"大型反应器通常采用抛光不锈钢内壁。"注意"be of"的用法。

⑤ "The problem is often overcome by the use of chilled water or the fitting of a reflux condenser which, via the continued refluxing of vinyl chloride monomer, utilises its latent heat of vapourisation for cooling purposes." 译为:"解决问题的方法是采用冷冻水或安装回流冷凝器,回流冷凝器连续回流氯乙烯单体,利用单体的蒸发潜热冷却反应物料。""utilises"与"vapourisation"均为英国拼法。

⑥ "heat-up of the reactor contents commenced." 译为:"开始反应器物料的加热。"

⑦ "The trend is towards the operation of large closed reactors which are only opened for maintenance or, possibly, occasional cleaning operations." 译为:"反应器的大型化封闭式操作是发展方向,仅仅在维护或偶然的清理时才打开聚合釜。"注意句型"The trend is towards…"的用法。插入语"possibly"强调不确定性。在一般情况下"which"前应加逗号,否则用"that"。

Exercises

1. *Complete the following blanks according to the text*

   It may be found that the polymer is insoluble in the monomer-solvent mixture from which it is formed. Polypropylene and PVC are two examples where the polymer has very limited _____ in the monomer. As polymerization proceeds, the polymer will _____ from the reacting mass to form a _____ phase of polymer swollen with the monomer-solvent mixture. This is called a _____ polymerization. (phase inversion can occur at high _____ to give a bulk polymerization.) A typical slurry polymerization is autorefrigerated. The heat of polymerization causes the reacting mass to _____; it is condensed and returned to reactor. The _____ processes for polyethylene and _____ are conceptually similar to slurry polymerizations. The _____ continuous phase is now a gas and the dispersed phase is a fluidized solid, but the heat of polymerization is still removed through the low-viscosity, continuous phase.

2. *Put the following words into English*

   悬浮聚合　分散剂　液滴合并　保护胶体　聚氯乙烯　脉动速度　搅拌器　液滴破裂　湍流　沉淀聚合　全混反应器

3. *Put the following passage into Chinese*

   Shear work is of importance in designing mixers for suspension polymerizations. Shear work is related to the number of times that a particle passes through a high shear region per unit time. In general the less flow in vessel, the less shear work for the same impeller configuration. At low process flow rates average residence time of a particle in a particular section of a tank varies more than with a high flow rate. Therefore, a minimum flow per unit volume is required to maintain a good particle residence time everywhere in the vessel. The high flow per unit volume in vessel is also required so that most of the particles in the tank will see approximately the same number of shear regions per unit time. Conversely, too much flow in a vessel will cause particle to be so small that micro-level shear rate assumes importance instead of macro-level shear rates.

Reading Materials

Agitation Affecting the Particle Properties of Suspension PVC

Agitation is of fundamental importance in the suspension PVC process. Together with the dispersant system it governs the stability of the suspension during polymerization and the particle size of the product formed. Agitation can also have an important role in determining other product properties such as porosity and bulk density.

In a liquid-liquid dispersion of low oil-to-water phase ratio, such as that existing in suspension polymerization systems, the individual droplets collide continuously with each other. Only some of these collisions result in immediate droplet coalescence however, as generally the droplets rebound owing to the elastic properties of the liquid film entrapped between them strengthened by the presence of the primary dispersant. Studies of droplet coalescence in the presence of protective colloids have found that there is a minimum contact time during which the chance of coalescence is negligible, but after this period of time, the probability of droplet coalescence increases with time as a result of thinning of the absorbed interfacial film.

In an agitated dispersion, the turbulent velocity fluctuations throughout the mixing tank tend to bring about separation of all adhering droplet clusters during the critical time period, so that the chance of coalescence is minimal. Both the force of adhesion between droplets and the forces caused by agitation (which tend to separate the droplets) depend on the diameter of the droplet. However, the chances of separation are greater the larger the droplets and thus, for a given level of agitation, there exists a minimum droplet diameter above which stabilization by agitation becomes possible. By contrast, the susceptibility towards droplet break-up, caused by local velocity fluctuations and shear forces near the impeller, is increased by increases in droplet size. The minimum droplet size that is able to remain stable to break-up will, of course, increase with decreasing agitator speed. However, in the case of very low agitation intensity large droplets may become unstable due to phase separation caused by specific gravity differences.

A detailed analysis of the role of agitation in stabilizing liquid-liquid dispersion is given by Church and Shinner. In this paper three equations are presented relating droplet diameter, in turn, to the forces causing droplet coalescence, break-up and phase separation. These three equations are plotted in Fig. 15.1 in which the agitation forces are represented by changes in stirrer speed.

A stable suspension is defined by the virtual absence of droplet break-up and coalescence occurring after a steady state has been reached. If the level of agitation required to separate two adhering droplets is higher than that required to produce instability in a single droplet, break-up and subsequent coalescence will occur continuously. It can be seen from Fig. 15.1 that stabilization by turbulence is only possible if the minimum diameter that can be stabilized by the turbulent pressure fluctuations is smaller than the maximum stable diameter for break-up and for suspension, as is the case within the triangle area illustrated. The relative positions of the lines representing the three equations would be expected to change with the type and level of the dispersant system being used, but the graph clearly predicts the existence of both a minimum and maximum level of agitation beyond which instability would occur.

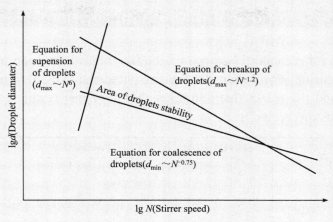

Fig. 15.1

Graphical representation of theoretical equations for break-up, coalescence and suspension of droplets in a stirred tank.

It is only much more recently, however, that papers have been published that quite clearly demonstrate the consequences of over-agitation in the suspension PVC process. Barclay qualitatively describes a minimum particle size obtained with increasing agitator speed beyond which product coarsening occurs. At about the same time Hofmann and Kummert presented experimental results showing the same effect.

——Butters G, ed. Particulate Nature of PVC. London: Applied Science Pub. Ltd., 1982, 17-19

Words and Expressions

collide	[kəˈlaid]	v.	碰撞
coalescence	[ˌkəuəˈlesns]	n.	合并
rebound	[riˈbaund]	n.; v.	回弹
entrap	[inˈtræp]	v.	诱陷
turbulent	[ˌtəːbjulənt]	a.	湍流的
fluctuation	[flʌkjuˈeiʃən]	n.	波动,脉动
adhere	[ədˈhir]	v.	黏附
stabilization	[steibəlaiˈzeiʃən]	n.	稳定
susceptibility	[səseptiˈbiliti]	n.	感受性
impeller	[imˈpelə]	n.	搅拌桨
break-up	[breikʌp]	v.	分裂
virtual	[ˈvəːtʃuəl]	a.	实质的
consequences	[ˈkɔnsiˌkwens]	n.	结果
qualitative	[ˈkwɔliˌteitiv]	a.	质量上的,定性的
coarsen	[kɔːsn]	v.	变粗糙
have an important role in			…起重要作用
bring about			引起
specific gravity			相对密度
by contrast			和…形成对照

UNIT 16 Styrene-Butadiene Copolymer

The synthetic rubber industry, based on the free-radical emulsion process, was created almost overnight during World War Ⅱ. Styrene-butadiene (GR-S) rubber created at that time gives such good tire treads that natural rubber has never regained this market.[①]

The GR-S standard recipe is

component	parts by weight	component	parts by weight
butadiene	75	potassium peroxydisulfate	0.3
styrene	25	soap flakes	5.0
n-dodecyl mercaptan	0.5	water	180

This mixture is heated with stirring and at 50℃ gives conversions of 5%～6% per hour. Polymerization is terminated at 70%～75% conversion by addition of a "short-stop", such as hydroquinone (approximately 0.1 part), to quench radicals and prevent excessive branching and microgel formation. Unreacted butadiene is removed by flash distillation, and styrene by steam-stripping in a column. After addition of an antioxidant, such as N-phenyl-β-naphthylamine (PBNA) (1.25 parts), the latex is coagulated by the addition of brine, followed by dilute sulfuric acid or aluminum sulfate. The coagulated crumb is washed, dried, and baled for shipment.

This procedure is still the basis for emulsion polymerization today. An important improvement is continuous processing illustrated in Fig. 16.1; computer modeling has also been described.

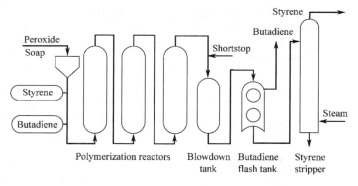

Fig. 16.1

SBR plant flow diagram courtesy of *hydrocarbon processing and petroleum refiner*

In the continuous process, styrene, butadiene, soap, initiator, and activator (an auxiliary initiating agent) are pumped continuously from storage tanks through a series of agitated reactors at such a rate that the desired degree of conversion is reached at the last reactor.[②] Shortstop is added, the latex warmed with steam, and the unreacted butadiene flashed off. Excess styrene is steam-stripped, and the latex finished as shown in Fig. 16.1.

SBR prepared from the original GR-S recipe is often called hot rubber; cold rubber is made at 5℃ by using a more active initiator system. Typical recipes are given in Table 16.1. At 5℃, 60% conversion to polymer occurs in 12~15h.

Table 16.1 Typical formulations for cold SBR

component	formulation 1	formulation 2
butadiene	72	71
styrene	28	29
tert-dodecyl mercaptan	0.2	0.18
diisopropylbenzene monohydroperoxide	0.08	
p-menthane hydroperoxide		0.08
ferrous sulfate heptahydrate	0.14	0.03
potassium pyrophosphate	0.18	
trisodium phosphate decahydrate		0.5
tetrasodium ethylenediaminetetraacetate		0.035
sodium formaldehyde sulfoxylate		0.08
rosin acid soap	4.0	4.5
water	180	200

Cold SBR tire treads are superior to those of hot SBR. Polymers with abnormally high molecular weight (and consequently too tough to process by ordinary factory equipment) can be processed after the addition of up to 50 parts of petroleum-base oils per hundred parts of rubber (phr). These oil extenders make the rubbers more processible at lower cost and with little sacrifice in properties; they are usually emulsified and blended with the latex before coagulation.

Recent trends have been toward products designed for specific uses. The color of SBR, which is important in many nontire uses, has been improved by the use of lighter-colored soaps, shortstops, antioxidants, and extending oils. For example, dithiocarbamates are substituted for hydroquinone as shortstop; the latter is used on hot SBR where dark color is not objectionable. A shortstop such as sodium dimethyldithiocarbamate is more effective in terminating radicals and destroying peroxides at the lower temperatures employed for the cold rubbers.

Free-radical dissociative initiators that function by dissociation of a molecule or ion into two radical species are normally limited to inorganic persulfates in the case of butadiene polymerization.

The other important class of free-radical initiators, redox systems, contain two or more components that react to produce free radicals. Dodecyl mercaptan added to control molecular weight also appears to aid free-radical formation by reaction with persulfate. The commercial importance of such chain-transfer agents or modifiers cannot be overemphasized. ③ Without molecular weight control the rubbers would be too tough to process.

——Tate D P, Bechea T W. Encyclopedia of Polymer Science and Engineering, 2nd ed. Vol. 2. Editor-in-chilf Kroschwitz JI. New York: John wiley & Sons, 1985. 553-555

Words and Expressions

tread	[tred]	*n.*; *v.*	踏，踩，外胎（面）
recipe	['resəpi]	*n.*	配方
n-dodecyl mercaptan			正十二烷基硫醇
potassium peroxydisulfate			过硫酸钾
flake	[fleik]	*n.*	薄片

short-stop		n.	终止剂
hydroquinone	['haidrəukwi'nəun]	n.	对苯二酚
quench	[kwentʃ]	v.	熄灭，抑制
flash distillation		n.	闪蒸
antioxidant	['ænti'ɔksidənt]	n.	抗氧剂
phenyl	['fenil]	n.	苯基
naphthylamine	['næfθilə'miːn]	n.	萘胺
latex	['leiteks]	n.	胶乳，乳液
coagulate	[kəu'ægjuleit]	v.	凝聚
brine	[brain]	n.	盐水
sulfuric acid		n.	硫酸
crumb	[krʌm]	n.	碎屑，团粒
bale	[beil]	v.	打包
abnormal	[æb'nɔːməl]	a.	反常的
extender	[iks'tendə]	n.	添加物〔剂〕，增量剂
sacrifice	['sækrifais]	v.	牺牲，损失
lighter-colored			浅色的
dithiocarbamate	[dai,θaiə'kɑːbəmeit]	n.	二硫代氨基甲酸盐
chain-transfer agent			链转移剂
objectionable	[əb'dʒekʃənəbl]	a.	反对的，使人不愉快的
dimethyldithiocarbamate		n.	二甲基二硫代氨基甲酸盐
sodium	['səudjəm]	n.	钠
tert-dodecyl mercaptan		n.	特十二烷基硫醇
diisopropylbenzene		n.	二异丙苯
ferrous sulfate heptahydrate		n.	硫酸亚铁七水合物
pyrophosphate	[,paiərəu'sʌlfeit]	n.	焦磷酸盐
aluminum sulfate		n.	硫酸铝
ethylenediaminetetraacetate		n.	乙二胺四乙酸盐
deca-	['dekə]		［构］十
tetra-	[tetrə]		［构］四
formaldehyde	[fɔː'mældi,haid]	n.	甲醛
sulfoxylate	[,sʌlf'ɔksileit]	n.	次硫酸盐
rosin acid		n.	松香酸
dissociation	[disəuʃi'eiʃən]	n.	分裂，分离，离解
persulfate	[pə'sʌlfeit]	n.	过硫酸盐
menthane	['menθein]	n.	薄荷烷（比较：methane 甲烷）
redox	['riːdɔks]	n.	氧化还原作用

Phrases

courtesy of　取自…（引用文献时）
followed by…　继之以，后面是，其次是

be superior to　胜过，优于，比…好
be substituted for N　代替了 N

Notes

① "Styrene-butadiene (GR-S) rubber created at that time gives such good tire treads that natural rubber has never regained this market." 译为："那时，丁苯橡胶制造的轮胎性能相当优异，使天然橡胶在市场上黯然失色。"

② "In the continuous process, styrene, butadiene, soap, initiator, and activator (an auxiliary initiating agent) are pumped continuously from storage tanks through a series of agitated reactors at such a rate that the desired degree of conversion is reached at the last reactor." 译为："在连续过程中，苯乙烯、丁二烯、皂、引发剂及活化剂（辅助引发剂）从贮槽连续泵送到搅拌釜组，泵送流率根据末釜的转化率控制。"

③ "The commercial importance of such chain-transfer agents or modifiers cannot be overemphasized." 译为："这种链转移剂具有极其重要的商品生产价值。"注意句型："cannot be overemphasized" 再强调也不过分。

Exercises

1. *Complete the following blanks according to the text*

   In _____ polymerization, which is applicable when the end-product is desired as a latex, monomer is dispersed by vigorous _____ in an immiscible liquid, usually water. Droplet size normally ranges from 0.1 to 1.0 microns. Emulsion stability, in the absence of agitation, is achieved by means of sufficient amounts of _____ and surfactants.

   Products made by this method include polyvinyl acetate for paint and _____; carboxylated styrene-butadiene copolymer; elastomers, such as _____ (SBR) or buna N (butadiene-acrylonitrile) rubbers, and _____ (ABS) polymer.

2. *Put the following words into English*

   乳化　凝聚　配方　胶乳　丁二烯　丙烯腈　十二烷基硫醇　终止剂　氧化还原　过硫酸钾　抗氧剂　合成橡胶

3. *Put the following passage into Chinese*

   Both conventional and inverse emulsion polymerizations comprise the emulsification of an immiscible monomer in a continuous medium followed by polymerization with a free radical initiator to give a colloidal sol of polymer particles. Both processes give "emulsion polymerization kinetics," i.e., a proportionality of both the polymerization rate and polymer molecular weight observed for mass, solution, and suspension polymerizations. The emulsion polymerization process can be divided into particle nucleation and particle growth stages and can be carried out using batch, semicontinuous, or continuous processes. Seed emulsion polymerization can be used to avoid the particle nucleation stage in all three processes. The many mechanisms proposed for the initiation of emulsion polymerization can be divided into four categories according to the locus of particle initiation: (1) monomer-swollen micelles; (2) absorbed emulsifier layer; (3) aqueous phase; (4) monomer droplets. These general principles are applied to: (1) the preparation of monodisperse latexes by seeded emulsion polymerization; (2) the locus of particle initiation for various monomers and initiators; (3) emulsion copolymerization; (4) core-shell emulsion polymerization; (5) polymerization in fine monomer droplets; (6) inverse emulsion polymerization.

Reading Materials

Steady-State Multiplicity in Continuous Emulsion Polymerization

The phenomenon of multiple steady states is seen in emulsion polymerization. Fig. 16.2 is a plot of steady-state monomer conversion as a function of reactor residence time for methyl methacrylate emulsion polymerization in a CSTR. A region of multiplicity is indicated by the fact that the upper and lower branches of the curve overlap between residence times of 30 and 50 minutes. The dotted line is an estimate of the shape of the unstable middle branch which is experimentally unobservable. The dashed lines indicate experimental instances of ignition and extinction. At 50 minutes residence time the system has been observed to move from the lower steady state of 54% conversion to the upper steady state at approximately 80% with no discernible change in operating conditions (ignition). Extinction has been observed when the residence time is changed from 30 minutes to 20 minutes on the upper branch, resulting in a drop in conversion from the upper to the lower steady-state values. The phenomenon of multiple steady states arises in emulsion polymerization for much the same reason as it appears in solution polymerization: the autocatalytic nature of the polymerization (due to the gel effect), combined with the mass balance, results in the possibility of steady-state multiplicity.

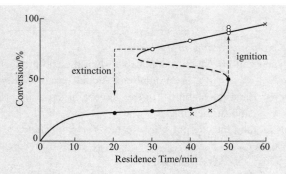

Fig. 16.2

Steady-state multiplicity in continuous emulsion polymerization of methylmethacrylate

Steady-state multiplicity can be an operational problem for a number of reasons. If one wishes to operate at an intermediate level of monomer conversion (perhaps to minimize viscosity or prevent excessive chain branching), one may be forced to operate in the unstable region, relying on closed-loop control to stabilize the operating point. This is tricky at best. Additionally, the steady state (upper or lower) to which the system goes on start-up will depend on how the start-up is effected. A careful start-up policy may be needed to assure that the system arrives at the desired steady state. In general, a conservative start-up, with the temperature and initiator concentration brought to steady-state values slowly will result in operation on the lower branch, while aggressive start-up (high temperature and/or high initiator concentration during start-up) will result in steady-state operation on the upper branch. Finally, large upsets in the process may cause ignition or extinction. This may lead to loss of temperature control in the case of ignition, or loss of reactor productivity in the case of extinction. A system designed to operate at the upper steady state will be operating way below design product yield at the lower steady state. Additionally, the product quality (MWD, CCD, etc.) will be different for the two operating points. The polymerization reactor designer should be aware of the potential for multiplicity, and, if possible, design the system to operate outside this region.

CSTR polymerization reactors can also be subject to oscillatory behavior. A nonisothermal CSTR free radical solution polymerization can exhibit damped oscillatory approach to a steady state, unstable (growing) oscillations upon disturbance, and stable (limit cycle) oscillations in which the system never reaches steady state, and never goes unstable, but continues to oscillate with a fixed period and amplitude. However, these phenomena are more commonly observed in emulsion polymerization.

High-volume products such as styrene-butadiene rubber (SBR) often are produced by continuous emulsion polymerization. This is most often done in a train of 5~15 CSTRs in series. Sustained oscillations (limit cycles) in conversion, particle number, and free emulsifier concentration gave been reported, under isothermal conditions in continuous emulsion polymerization systems. This limit cycle behavior leaves its mark on the product in the form of disturbances in the molecular weight distribution and particle size distribution which cannot be blended away. Fig. 16.3 shows evidence of a sustained oscillation (limit cycle) during emulsion polymerizaion of methyl methacrylate in a single CSTR. Comparison of the monomer conversion and surface tension data graphically illustrates the mechanism

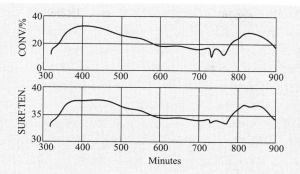

Fig. 16. 3

Limit cycle oscillation in continuous emulsion polymerization of methyl methacrylate

of oscillation. It will be noted that the surface tension oscillates with the same period as the conversion (6~7 residence times). This can be explained with the classical micellar initiation mechanism (or with homogeneous nucleation). Beginning at a time of about 300 minutes, the conversion rises rapidly as new particles form and old particles grow. As the particle surface area increases, additional surfactant is adsorbed on the particles. Meanwhile micelles dissociate to keep the aqueous phase saturated. Once all of the micelles have dissociated, it is no longer possible to maintain the aqueous phase at saturation, and the surface tension begins to rise. This is observed at about 320 minutes. At the point at which micelles are no longer present, micellar initiation stops and the rate of polymerization slows. Eventually, since particles are washing out while no new particles are being formed, the conversion begins to fall. Since the total particle surface area is decreasing at this point, and since surfactant is continually being introduced with the feed, the surface tension falls as the aqueous phase reapproaches saturation. As the aqueous phase becomes saturted initiation begins again. Saturation of aqueous phase may be observed by noting the point at which the surface tension reaches its CMC value. As new micelles are formed they adsorb free radicals, become polymer particles, and begin to grow and adsorb surfactant. The cycle then repeats.

Modeling studies show that while the instability arises above the CMC (and is promoted by large values of initiator concentration and residence time, and low surfactant concentration), it is the on/off nature of the nucleation mechanism which governs the nature of oscillations in monomer conversion. The surface tension oscillation leads the conversion oscillation by approximately one residence time. This is consistent with the above explanation since changes in surfactant concentration are quite rapid while changes in the number of particles and rate of reaction require a finite growth time to appear as changes in the monomer conversion. Damped oscillations at start-up have been noted for a large number of monomer systems.

Damped oscillations will result in lost productivity since the product during these transients may be off quality. Unstable oscillations will, of course, preclude continued operation. Limit cycle oscillations, while not unstable, will result in a product having a quality (MWD, CCD, etc.) which varies with time in a cyclic fashion. In most cases this is undesirable. As in the case of multiplicity, the polymerization reactor designer must be aware of the potential for oscillatory phenomena, and should attempt to specify operating conditions at which these phenomena do not exist. In emulsion polymerizations, oscillations (both damped and sustained) are undesirable since the product is not of a consistent quality, and oscillations in free surfactant concentration may induce coagulation and reactor fouling. Several methods of eliminating oscillations in emulsion polymerization have been suggested. Poehlein has used a plug flow reactor upstream of a CSTR train. All polymer particles are nucleated in the PFR. Since PFR kinetics are essentially those of a batch reactor (and such oscillations do not occur in batch reactors), no oscillations occur. The CSTRs, then, are used to grow the existing particles. By segregating particle nucleation from particle growth, oscillations are eliminated. Another approach has been taken by Penlidis and others. This involves using a small CSTR as a seeder reactor. All polymer particles are formed in the seeder. A portion of the monomer and water is bypassed around the seeder in such a way as to dilute out any remaining micelles in the reactor immediately following the seeder. Once again, nucleation and growth have been segregated, and oscillations are eliminated.

——Schork F J. Reactor Operation and Control. in: McGreavy C, ed. Polymer Reactor Engineering. New York: Blackie Academic & Professional, 1994. 161-164

Words and Expressions

multiplicity	[,mʌlti'plisiti]	n.	多样性
steady-state	['stedi-steit]	a.	不变的
overlap	[,əuv'læp]	v.	与…交[重]叠
ignition	[ig'niʃən]	n.	点火
extinction	[iks'tiŋkʃən]	n.	消灭, 熄火
discernible	[di'sə:nəbl]	a.	可辨别的
autocatalytic	[,ɔ:tou,kætə'litik]	a.	自催化的
rely	[ri'lai]	v.	依赖
tricky	['triki]	a.	狡猾的
oscillatory	['ɔsileitəri]	a.	振动的
damp	[dæmp]	v.	阻尼
micellar	[mai'selə]	n.	胶束的
reapproach	[riə'prəutʃ]	v.	再接近, 再接洽
dotted line			点线
dashed lines			虚线
conservative start-up			保守开车法
aggressive start-up			进取开车法
loss of temperature control			温度失控
MWD, molecular weight distribution			分子量分布
CCD, copolymer composition distribution			共聚合组成分布
high-volume products			大吨位产品
off quality			质量低劣

UNIT 17　Heat Transfer Process

The formation of polyvinyl chloride is an exothermic reaction. One of the functions of the reactor vessel is to provide the cooling surface for heat removal. The reactor size is limited by the ratio of reactor volume to cooling surface and the agitation system needed to maintain the monomer-water mixture at the proper droplet size. Commercial reactors of 25000 to 50000gal capacity are standard for suspension polymerization industry. The reactor productivity will depend on the heat transfer capacity of the vessel. The higher the heat transfer capacity, the more polymer that can be made per unit of time.

1. Reflux condenser utilization

A reflux condenser functions by condensing vapor to liquid and removing the latent heat of vaporization from the system. In large reactors, a reflux condenser is the most effective way to increase the available cooling surface without increasing the length/diameter ratio of the reaction vessel. The use of a reflux condenser to provide additional cooling capacity in vinyl chloride polymerization reactors was discussed by Terwiesch [1976] for reactor of 50000gal capacity. A Conoco [U. S. Pat. 3980626, 1976] patent describes the installation and use of a reflux condenser. Shinetsu [U. S. Pat. 4136242, 1979] indicates that the reflux condenser should be used only after 5% of the monomer is converted to polymer, because utilization of the reflux condenser before 5% conversion results in extensive reflux condenser fouling and a coarser resin.

The reflux condenser operation envelope (amount of reflux, start-up time for reflux) will depend on the ability of the reactor agitation system to reincorporate the refluxed monomer effectively into the polymerizing mass, and on the foaming tendency of the polymerization recipe.① Venting procedures have been described to remove the noncondensable gases from the vapor phase of the reactor and reflux condenser. The level of noncondensable gases in the reactor vapor phase will be affected by the quality of the monomer, how well the reactor has been evacuated prior to polymerization, and whether the polymerization process generates an inert gas. The use of azo initiator will result in the formation of nitrogen as a by-product of the initiator decomposition. In addition, a carbonate buffer system will produce CO_2 gas if the aqueous phase becomes acidic. Both situations will result in the generation of inert, noncondensable gases which will reduce the efficiency of the reflux condenser.

2. Reactor jacket operation

The reactor jacket serves the dual purpose of the first heating the reaction mixture to polymerization temperature by passing a high-temperature fluid through the jacket and then maintaining polymer slurry at a fixed temperature by passing a low-temperature fluid through the jacket. In large reactors additional cooling is available through the use of a reflux condenser. In a batch polymerization process the reaction mass is heated as quickly as possible to the polymerization temperature so as to increase the productivity of the reactor. This requires that steam-heated water at about 180 to 190°F be circulated in the jacket. It has been suspected that this type of heat-up process leads to increased wall fouling. ② An alternative approach would be to use preheated process water. The second alternative would be to sparge steam directly into the reactor. Direct sparging into a reactor has many problems associated with product contamination and is not widely practiced. A third approach is to heat the reactor adiabatically by means of a very active initiator system.

3. Agitation system operation

The suspension droplet is an easily deformed elastic body in constant motion within the reactor. The agitation system, which is composed of rotating agitator and fixed baffles, serves to ensure steady heat transfer through mixing and to develop and maintain the size distribution of the monomer droplet in the water. As a trend it has been observed that structures within the reactor which lead to fluid dynamic flow turbulence will increase reactor wall fouling. ③ Therefore, baffle support braces need to be designed for the optimum fluid dynamic stability.

——Nass L I, Heiberger C A, ed. Encyclopedia of PVC, 2nd Edition. New York: Marcel Dekker, Inc., 1986. 87-90

Words and Expressions

heat transfer			传热
reflux condenser			回流冷凝器
gal	['gæl]	n.	gallon 的缩写，(体积) 加仑
vaporization	[veipərai'zeiʃən]	n.	蒸发
resin	['rezin]	n.	树脂
envelope	['envə,ləup]	n.	框（图）
reincorporate	[ri:in'kɔ:pə,reit]	v.	再合并
foam	[fəum]	v.	发泡
noncondensable	[nɔnkən'densəbl]	a.	不可冷凝的
inert	[i'nə:rt]	a.	惰性的
azo	['æzəu]	n.	偶氮基
by-product	['bai'prɔdʌkt]	n.	副产物
decomposition	[di:,kɔmpə'ziʃən]	n.	分解
dual	['dju:əl]	a.	双重的
carbonate	['kɑ:bəneit]	n.	碳酸盐
buffer	['bʌfə]	n.	(pH) 缓冲液
heat-up	[hi:tʌp]	v.	加热升温
sparge	[spɑ:dʒ]	v.	喷射
contamination	[kəntæmi,neiʃən]	n.	污染
adiabatical	['ædiəbætikəl]	a.	绝热的
deformed	[di'fɔ:md]	a.	不成形的，残废的，变形的
baffle	['bæfl]	n.	挡板
brace	[breis]	n.; v.	支柱

Phrases

by means of　依靠，根据　　　　　　　　　　有可能…
It has been suspected that…　怀疑，推测，估计，

Notes

① "The reflux condenser operation envelope (amount of reflux, startup time for reflux) will depend on the ability of the reactor agitation system to reincorportate the refluxed monomer effectively into the polymerizing mass, and on the foaming tendency of the polymerization recipe." 译为："回流冷凝器的操作

（回流量，开始回流时间）取决于反应器搅拌机构有效再分散回流单体的能力和聚合配方的发泡性。"

② "It has been suspected that this type of heat-up process leads to increased wall fouling." 译为："这种加热升温方法可能使粘壁物增加。"注意句型"It has been suspected that…"。

③ "As a trend it has been observed that structures within the reactor which lead to fluid dynamic flow turbulence will increase reactor wall fouling." 译为："反应器内的这种（挡板）结构能促进湍流但使粘壁物增加。"

Exercises

1. *Complete the following blanks according to the text*

   Suspension polymerization reactors are generally vertical, agitated _____ . One of their essential function is _____ the heat of polymerization, which can be appreciable. For PVC, it is approximately 1513 kJ/kg. To remove this heat, which is released over the relatively short period of 5 to 10 hr., reactor are generally _____ and water cooled. Peak heat release may be several times larger than the _____ .

   Overall heat transfer coefficients vary with types of reactors, which are usually constructed either of stainless steel or glass-lined carbon steel. The overall heat transfer coefficients for a _____ vessel are 312 to 397 W/ (K·m^2), whereas for a _____ unit they can be as high as 125. The difference between the two is due to the additional heat transfer resistance offered by the glass layer.

   As reactors are scaled up in size, problems are generally encountered with _____ surfaces. For a cylindrical vessel, the straight-side, jacket heat transfer area does not _____ proportionately to the reactor volume when dimensional similarity is maintained.

2. *Put the following words into English*

   放热反应　夹套　回流冷凝器　传热　聚合物淤浆　内冷管　绝热反应器　挡板　温度失控　多稳态　等温聚合

3. *Put the following passage into Chinese*

   Some polymer processes are carried out in solvent at relatively low final polymer concentrations, either for processing ease or to achieve the desired final polymer properties. Examples are certain *cis*-butadiene rubber processes, where the volume of the diluting solvent is sufficiently high that cooling may be achieved by feeding cold diluent solvent to the reactor. In some processes, this is the only cooling means, but in many others, the design engineer should not overlook the technique as a significant means of providing additional reactor cooling.

   For example, consider a hypothetical case where the final polymer concentration in solution is 10%. Assuming a latent heat of polymerization of 698 kJ/kg. and a solvent specific heat of 2093 J/ (kg·k), feeding solvent to the reactor at a temperature 67 degree. below the final reaction temperature is sufficient to remove all of the heat of polymerization.

   The blending rate of the cold solvent in the polymer solution should be carefully considered because it is desirable to maintain as uniform a temperature distribution as possible to avoid variations in product properties. Although this is not difficult to achieve in pilot plants, it may become a severe problem in commercial scale. The ability to blend cold-solvent feed into the polymer solution may place practical restriction on the degree to which the solvent can be subcooled below reaction temperature. This is not for concern, though, if the reactor agitator can be operated in the turbulent range.

Reading Materials

CSTR Heat Balance Dynamic Behavior

By the dynamic behavior of a polymerization reactor is meant the time evolution of the states of the reactor. The states are those fundamental dependent quantities which describe the natural states of the system. A set of equations which describes how the natural state of the system varies with time is called the set of state equations. Temperature, pressure, monomer conversion and copolymer composition could be considered states of a polymerization reactor. Independent variables such as coolant temperature in jacketed reactor or initiator addition rate are not states but (controlled or uncontrolled) inputs. For various reactor types, different modes of dynamic behavior are observed. These can range from stable operation at a single steady state to instability, multiple steady states or sustained oscillations.

The widest spectrum of dynamic behavior is observed in the CSTR. As we have seen, the use of a CSTR of CSTR train for polymerization reactions may be justified in some cases by kinetic considerations. However, before implementing CSTR polymerization, the engineer should be aware of the unique dynamics associated reactions in a CSTR which are exothermic and/or autocatalytic, or involve, nucleation phenomena.

Consider an irreversible first-order exothermic reaction in a CSTR. The rate of thermal energy release by reaction can be plotted versus temperature, as shown by the curve Q_g in Fig. 17.1. At low temperature, the reaction rate is low, and the slope of Q_g is slight. At high temperatures the reactor is operating at a high level of conversion (low reactant concentration) and additional increases in temperature result in negligible increase in reaction rate and heat evolution. If the reactor is jacketed, the rate of heat removal (for fixed jacket temperature) is linear with reaction temperature. Thus, depending on operating conditions, the rate of heat removal may be represented by the various heat removal lines marked Q_r in Fig. 17.1. Since, at steady state, the rate of heat generation must equal the rate of heat removal, steady-state conditions can exist only at the intersection of the Q_g and Q_r curves. Depending on operating conditions (the slope and position of the Q_r line) there may be one or three steady states. In the case of three steady states, it easily may be seen that the upper and lower steady states are stable since perturbations in temperature will result in the system returning to its original position when the perturbation is removed. The middle steady state, however, can be seen to be unstable since any perturbation will drive the system away from the middle steady state and toward the upper or lower steady state (depending on the direction of the perturbation). This type of heat balance multiplicity is common in CSTR polymerization due to the highly exothermic nature of polymerization reactions. The presence of a gel effect will augment the potential for multiplicity.

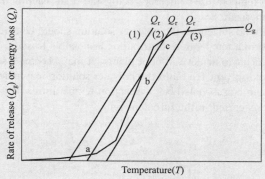

Fig. 17.1

Heat balance multiplicity during exothermic reaction in a CSTR

This phenomenon can be observed in free radical polymerization (nonisothermal) due to the exothermic nature of the polymerization reaction. However, due to the gel effect, it is also observed in some isothermal free radical polymerizations in a CSTR. Fig. 17.2 shows the rate of polymerization plotted versus monomer conversion for the free radical solution polymerization of methyl methacrylate. Unlike a more common reaction in which the rate of reaction falls monotonically with conversion, the rate of reaction rises with conversion due to the onset of the gel effect. Thus the system can be thought of as autocatalytic. At high conversions the polymerization becomes monomer-starved, and the rate of polymerization falls to zero. At a fixed residence time, there must be a specific rate of polymerization to produce a given monomer conversion. Thus the mass balance is represented by the dotted lines in Fig. 17.2. The slope of the mass balance line will vary with operating conditions, but it will always pass through the origin since at zero reaction rate the monomer conversion is zero. Inspection of Fig. 17.2 will reveal that for mass balances (operating lines) with slopes between the two dotted lines, three steady states will exist since an intersection of the reaction rate curve and the operating line defines a steady state. This may be better seen by referring to Fig. 17.3, where monomer conversion has been plotted versus reactor residence time. (A similar plot will result from the heat balance multiplicity in a nonisothermal CSTR.) It may be seen that over a range of residence times, three values of monomer conversion are possible. As before, the upper and lower steady states are usually stable, while the middle steady state is not.

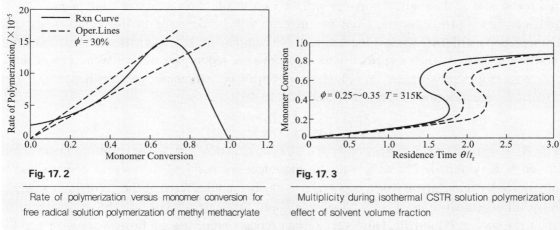

Fig. 17.2

Rate of polymerization versus monomer conversion for free radical solution polymerization of methyl methacrylate

Fig. 17.3

Multiplicity during isothermal CSTR solution polymerization effect of solvent volume fraction

——Schork F J. Reactor Operation and Control. in: McGreavyC, ed. Polymer Reactor Engineering. New York: Blackie Academic & Professional, 1994. 159-161

Words and Expressions

evolution	[i:və'lu:ʃən]	n.	进展，发展
spectrum	['spektrəm]	n.	光谱，频谱
implement	['implimənt]	v.	贯彻，实现
irreversible	[iri'və:rsəbl]	a.	不可逆的
intersection	[,intə'sekʃən]	n.	交叉点
perturbation	[pə:tə:'beiʃən]	n.	干扰
nonisothermal	[nɔn,aisə'θə:məl]	a.	非等温的
monotonical	[,mɔnə'tɔnikəl]	a.	单调的
onset	['ɔn,set]	n.	开始，发作
starve	[stɑ:v]	v.	饥饿
the set of state equation			状态方程组
gel effect			凝胶效应

UNIT 18 Copolymer Composition Distributions Affected by Micromixing

In multicomponent polymerizations by a free-radical mechanism, polymer molecules will usually have a composition different than that of the monomer mixture from which they were formed. The polymer will be rich in those monomers with high reaction rates and deficient in monomers with low rates. For copolymerizations, the instantaneous compositions are related by Equation (18.1).

$$\left(\frac{A}{B}\right)_{polymer} = \frac{dA}{dB} = \frac{A(A\gamma_A + B)}{B(A + \gamma_B B)} \tag{18.1}$$

where A and B are the monomer concentrations.

In batch or piston flow reactors, Equation (18.1) can be integrated to give monomer and polymer compositions as a function of overall conversion to polymer. The composition of the polymer first formed is given by Equation (18.1) with initial values A_0 and B_0 substituted for A and B. The initial polymer will be rich in the more reactive component, but as polymerization proceeds, the monomer mixture will be depleted in this component, and compositions will drift toward the less reactive component. If a copolymer of uniform composition is desired, the more reactive monomer must be fed to the reactor as a function of time.

Azeotropes exist whenever γ_A and γ_B are both less than one or greater than one.① The azeotropic composition is given by Equation (18.2)

$$\left(\frac{A}{B}\right)_{azeo} = \frac{1-\gamma_B}{1-\gamma_A} \tag{18.2}$$

For A = styrene and B = acrylonitrile, γ_A = 0.41 and γ_B = 0.4. The azeotrope lies at about 38 mol% acrylonitrile (24 wt%); at this monomer composition, the polymer composition is identical. A chemically uniform copolymer is easily manufactured at the azeotrope; indeed, large tonnage quantities of SAN copolymer② and the matrix phase of ABS are made at or about 24 wt% acrylonitrile. However, certain product properties are favored at higher acrylonitrile contents, and commercial products are also made with average compositions of 28 to 36 wt% acrylonitrile. Close control of the distribution about the average composition is needed to ensure mechanical compatibility and good color of the polymer.

A perfect mixer gives chemically uniform copolymer. To obtain 28 wt% acrylonitrile in the polymer, the reacting mixture should contain about 36 wt% acrylonitrile according to Equation (18.1); feed composition to the reactor can be adjusted to achieve this. For other reactor types, some spread in copolymer composition distribution must be accepted.

Fig. 18.1 shows the effects of both macromixing and micromixing on the styrene-acrylonitrile system. For these results, the feed composition was selected to give an initial polymer composition of 28 wt% acrylonitrile, and the mean residence time in each type of reactor was adjusted to give 60% overall conversion. The level of macromixing was adjusted by varying the residence time distribution from that of a perfect mixer to that of a piston flow reactor.③ Intermediate residence time distributions correspond to a stirred tank in series with a piston flow reactor and are characterized by the dimensionless variance σ^2. ④ The magnitude of σ^2 represents the fraction of the system volume that is stirred. The perfect mixer gives a uniform composition distribution at the average value of 28% acrylonitrile while a batch reactor shows a spread with compositions ranging up to 31%. A segregated stirred tank gives the broadest copolymer composition distribution with the tail of the distribution approaching pure polyacrylonitrile. Intermediate results illustrate the effect of varying the level of macromixing at fixed levels of micromixing.

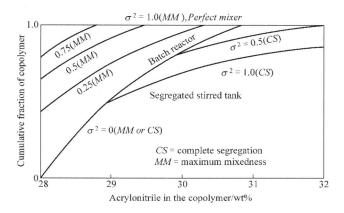

Fig. 18. 1

Effects of micromixing and macromixing on copolymer composition distributions

Other studies on copolymer composition distributions have also found perfect mixers and segregated stirred tanks to represent extremes within the normal mixing region of Fig. 18. 2. O'Driscol and Knorr (1969) treated the three idealized reactors for copolymer reactivity pairs of $\gamma_A = 20$, $\gamma_B = 0.015$ and $\gamma_A = 0.5$, $\gamma_B = 1$. Mechlenburgh (1970) also analyzes the three idealized reactors but extends the analysis to a segregated tubular reactor with a parabolic velocity distribution and to a piston flow reactor with backmixing by way of axial diffusion. Szabo and Nauman (1969) analyzed a tubular reactor with power law flow, Equation (18.3), in a recycle loop. At high-recycle ratios this reactor approaches the performance of a perfect mixer for the design of agitation systems.

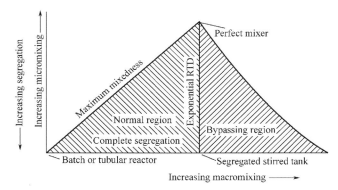

Fig. 18. 2

Mixing space for continuous flow reactors

$$V = V_0 \left[1 - \left(\frac{r}{R}\right)\right]^{n+1/n} \quad (18.3)$$

———Nauman E B. Synthesis and Reactor Design.
in: Baijal MD, ed. Plastics Polymer Science and Technology.
New York: John Wiley & Sons Inc, 1982. 508-509

Words and Expressions

multicomponent	[mʌltikəm'pəunənt]	n.	多组分[元]
deficient	[di'fiʃənt]	a.	缺乏的,不足的
instantaneous	[,instən'teinjəs]	a.	瞬时的
piston flow			活塞流
overall conversion		n.	总转化率
deplete	[di'pli:t]	v.	耗尽
azeotropic	[ə'zi:ətrəupik]	n.	共沸混合物
acrylonitrile	['ækriləunai'tril]	n.	丙烯腈
matrix	['meitriks]	n.	基体,基质;矩阵
close control			精密[严格]控制
mixer	['miksə]	n.	搅拌器
macromixing	['mækrəu,miksiŋ]	n.	宏观混合
micromixing	['maikrəu,miksiŋ]	n.	微观混合
mean residence time			平均停留时间
residence time distribution			停留时间分布
dimensionless	[di'menʃənlis]	n.	无量纲
variance	['vɛə:riəns]	n.	方差
parabolic	[,pærə'bɔlik]	a.	抛物线[面]的,比喻的
backmixing	['bækmiksiŋ]	n.	返混
approach	[ə'prəutʃ]	n.	逼近,方法

Phrases

drift toward 向…漂移	large tonnage quantities of 大吨位…
substituted for 代替	in series with… 和…串联

Notes

① "Azeotropes exist whenever γ_A and γ_B are both less than one or greater than one." 译为:"无论是 γ_A 和 γ_B 小于1或大于1均存在恒组成共聚物。"

② SAN copolymer,即 styrene-acrylonitrile copolymer,苯乙烯-丙烯腈共聚物。

③ "The level of macromixing was adjusted by varying the residence time distribution from that of a perfect mixer to that of a piston flow reactor." 译为:"宏观混合程度通过停留时间分布调节,范围从全混反应器到平推流反应器。"

④ "Intermediate residence time distributions correspond to a stirred tank in series with a piston flow reactor and are characterized by the dimensionless variance σ^2."

句中 "Intermediate residence time distributions" 意指 "介于全混釜和平推流反应器中间的停留时间分布"。"in series with…" 为一个介词短语,作后量定语,意指 "和…串联的…"。"a stirred tank in series with a piston flow reactor" 可译为 "由一个搅拌反应釜和一个平推流反应器串联而构成的反应器"。全句译文:"(介于全混釜和平推流反应器)中间的停留时间分布相当于由一个全混釜和一个平推流反应器串联而构成的反应器,(其停留时间分布)可用无量纲方差 σ^2 来表征"。

Exercises

1. *Complete the following according to the text*

   (1) If a copolymer of uniform composition is desired, we will _____ .

   (2) "Azeotrope" in Paragraph 3 means _____ .

(3) Fig. 16.1, the residence time distributions are characterized by the dimensionless variance σ^2, for a piston flow reactor, $\sigma^2 = $ _____ .

(4) The article aims to _____ .

2. *Put the following words into English*

共聚物组成分布　　　宏观混合　　　微观混合　　　平推流
理想混合　　　停留时间分布　　　返混　　　恒组成共聚物
无量纲方差　　　SAN 共聚物

3. *Put the following into Chinese*

Fig. 18.2 attempts a simplified graphical representation of the kinds of mixing possible in continuous flow reactors. Macromixing is one dimension of the two-dimensional mixing space. This dimension is characterized by the residence time distribution, and the dimensionless variance of the distribution σ^2 provides a possible—but not unique—quantitative measure of macromixing. Dimensionless variances range from zero to one in the normal region and correspond to residence time distributions ranging from the uniform distribution of piston flow to the exponential distribution of stirred tanks. Bypassing can give dimensionless variances greater than one; and there is no upper limit for macromixing when quantified in this fashion.

Micromixing is the other dimension depicted in Fig. 18.2, and one possible quantification of this is 1-J where J is the degree of segregation defined by Danckwerts (1958) and Zwietering (1959). The theoretical range for J is zero to one, but $J = 0$ is possible only with an exponential residence time distribution. All other distributions have some minimum possible J. Thus there are some maximum possible values for 1-J, and the mixing space shown in Fig. 18.2 is bounded from above.

The triangular region corresponding to normal mixing has three idealized reactors at its vertices, and it has long been tacitly assumed that these three reactors provide limits on the molecular weight distribution possible in real systems. Reactors with bypassing may in fact give distribution that fall outside these limits; but the batch reactor, segregated stirred tank and perfect mixer do represent realistic limits for normal mixing situations. Consequently, the theoretical molecular weight distributions for these idealized reactors merit detailed study.

Reading Materials

Micromixing

Even for isothermal reactions, the residence time distribution is adequate to predict yields only if the reaction is first order. For other types of reactions, further information on mixing patterns is needed for a precise yield prediction although relatively close bounds on the yield can be calculated from knowledge of the residence time distribution alone. The theory behind these statements is called "micromixing theory". It springs from the observations of Danckwerts and Zwietering that fixing the residence time distribution still leaves a degree of freedom on molecular level mixing possible within a reactor.

The residence time distribution merely defines what Levenspiel has termed "macromixing" and does not define the molecular scale mixing that results from diffusion and that is called micromixing. Macromixing refers to those gross flow processes that cause different fluid elements to have different residence times. These same flow processes may or may not cause such intimate comingling of materials that diffusion becomes important so that micromixing occurs. Thus the beads in a continuous suspension process can show a distribution of residence times without any mixing on the molecular scale. Such a system is said to be "completely segregated", and complete segregation represents one limit on micromixing, that where there is no molecular mixing at all. There is also an upper limit on micromixing that corresponds to the maximum amount of molecular level mixing possible with a given residence time distribution. Zwietering called this condition "maximum mixedness".

The state of complete segregation is easily imagined for any residence time distribution; simply consider the individual fluid elements to be encapsulated so that they experience the gross flow patterns of the vessel without any interchange of molecules. Alternatively, a completely segregated reactor can be modeled as a piston flow reactor with multiple side exits as shown in Fig. 18.3(a). These side exits can be so distributed as to match an arbitrary residence time distribution. From a yield viewpoint, this arrangement is equivalent to treating the system as a large number of piston flow elements in parallel and is thus a completely segregated reactor.

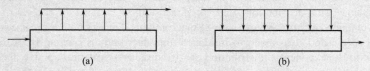

Fig. 18.3

Piston flow reactor models
(a) Completely segregated reactor; (b) Maximum mixdness reactor

The other limit of micromixing, namely the condition of maximum mixedness, seems harder to visualize. Zwietering, however, showed that it could be modeled as a piston flow reactor with multiple side entrances. See Fig. 18.3(b).

These two piston flow models have the same residence time distribution but in general will predict different reaction yields. In the completely segregated reactor, mixing between fluid elements that have spent different times in the vessel occurs as late as possible, namely, in the exit stream. In the maximum mixedness reactor, this mixing occurs as early as possible, namely, at the various entrance points.

Zwietering showed that complete segregation and maximum mixedness represent bounds on the amount of mixing possible between molecules that have spent different times in the reactor. They also provide absolute limits on conversion under rather broad conditions. Assuming premixed feed, Chauhan, Bell, and Adler have shown that complete segregation gives the highest possible conversion when the rate expression R_A is concave-up, $d^2R_A/dC_A^2 \geqslant 0$; and maximum mixedness gives the highest conversion when R_A is concave-down, $d^2R_A/dC_A^2 \leqslant 0$, if $d^2R_A/dC_A^2 = 0$ the rate expression is linear, the reaction is first order, and yield is independent of micromixing.

Novad and Thyn give generalized plots that show the complete segregation and maximum mixedness conversions for various reaction orders and residence time distributions. It turns out that the largest difference in conversion for a second-order reaction occurs in a stirred tank reactor and is only about 7%. This small difference explains the difficulties encountered by early workers who attempted to measure micromixing by performing second-order reactions in stirred tanks. The difficulty is aggravated by the fact that stirred tank reactors in the normal regime of turbulence and high diffusivities tend to be quite close to maximum mixedness. However, some degree of segregation can be found with unmixed feed, gentle agitation, or vary fast reactions. Departures in yield from that of a maximum mixedness reactor provide a measure of micromixing that in turn can be used to evaluate parameters in a model.

A complete model of an isothermal continuous flow reactor must obviously predict the extent of both macromixing and micromixing. The usual approach to this problem is to use any conventional submodel for the residence time distribution and to devote one additional parameter for predicting levels of micromixing possible with this residence time distribution. The simplest way to create such a composite macro/micromixing model is just to treat the system as two different reactors in parallel. The two reactors have identical residence time distribution, but one is completely segregated and the other is in maximum mixedness. The micromixing parameter represents the division of entering material between the two reactors. This model, which was suggested by Methot and Roy, can be considered part of a general class of "two-environment" models first discussed by N_y and Rippin and Weinstein and Adler. However, in the usual formulation "of the two-environment" model, the division between the segregated and maximum mixedness regions is based on the age of a fluid element. Typically, fluid enters the reactor in the segregated environment and leaves from the maximum mixedness environment. Chen and Fan have proposed a reversed two-environment model as more representative of polymer reactions. It is supposed that the reactants are initially well mixed but become segregated as viscosities increase.

There are other models of combined macromixing and micromixing that take a more mechanistic approach to mixing in turbulent flow. In the axial dispersion model, for example, the Peclet number simultaneously defines the residence time distribution and the degree of segregation. Within a limited range this model is also valid for laminar flow of non-Newtonian fluids. The model of Manning, Wolf, and Keairns is representative of those suitable for stirred tanks in the turbulent regime.

——Nauman E B. Synthesis and Reactor Design.
in: Baijal M D, ed. Plastics Polymer Science and Technology.
New York: John Wiley & Sons Inc., 1982. 491-492

Words and Expressions

comingle	[kɔ'miŋgl]	v.	混合，参合，commingle 的变体
encapsulate	[in'kæpsjuleit]	v.	胶囊化，压缩
arbitrary	['ɑ:bitrəri]	a.	武断的，任意的
visualize	['vizjuə,laiz]	v.	可视化，观察
generalize	['dʒenərə,laiz]	v.	归纳，使一般化
concave	[kɔn'keiv]	n; a.	凹
departure	[di'pɑ:tʃə]	n.	出发，偏离
simultaneous	[,siməl'teinjəs]	a.	同时存在或发生的
first-order reaction		n.	一级反应
spring from			起源于…
concave up			上凹
concave down			下凹
non-Newtonian fluid			非牛顿流体

UNIT 19 Introduction to Modelling of Polymerization Kinetics

The objective of a rational reactor design scheme is the development of a polymer reactor configuration optimal in some sense. The measures of product polymer properties typically available to the reactor engineer are the molecular weight distribution, expressed in terms of the distribution function itself or in terms of the moments of this distribution, and the composition and sequence distributions in the case of copolymers.

In this text are presented those mathematical techniques most useful in modelling polymerization reactors. The methods of interpretation and analysis are quite distinct but are in fact related. Fig. 19.1(a) gives the schematic interrelationships of the available techniques.

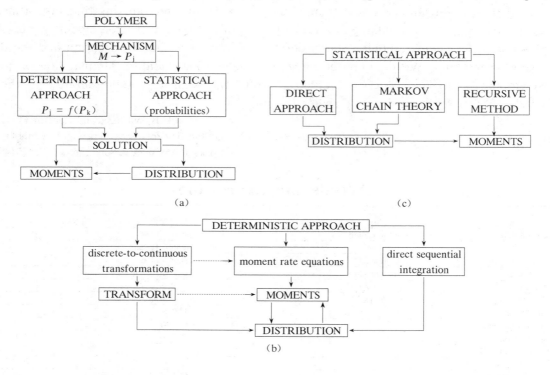

Fig. 19. 1

Statistical methods for describing polymer reactions

The first step in any analysis is the postulation of a kinetic mechanism which governs the set of reactions comprising the polymerizations. For a particular reactor type this mechanism may lead via the mass balance equations to a set governing the evolution of the concentration of polymer with a given chain length[①]. This opens several avenues by which the equations might be solved in order to glean that information necessary to understand the polymerization. Fig. 19.1(b) summarizes this.

The equations could be solved directly, usually by numerical means, to yield the entire distribution of polymer chain lengths. This procedure is quite tedious and lengthy, since the equations form an infinite set. Alternatively, moment equations can be developed from the original set of evolution equations. Inasmuch as colligative methods of analysis yield data related to the moments, it would appear that the solution of a truncated set of moment equations might suffice to give an adequate description of a polymerization. ②

The equations for the concentrations of polymer of chain length k can be transformed by a number of related methods, termed here transformation techniques, into a finite set whose solution yields both the moments of the polymer chain length distribution as well as the polymer chain length distribution (PCLD) itself.

Often these techniques require extensive use of computers or the best they can offer is moments of the distribution. ③ Due to these limitations it is, in some cases, efficient and/or informative to bypass the mass balance equations and develop a statistical or probabilistic approach. In fact, when the concept of polymerization reaction engineering started to develop around the 1940's, this was the predominant approach. Cases arise when the polymerization can easily be thought of as some kind of classical statistical process such as Markov or Poisson process. ④ Statistical approaches have the advantages of simplicity and directness in some cases, as well as occasionally revealing certain characteristics of the polymerization process which is marked by a purely deterministic treatment. A schematic of the various statistical methods is shown in Fig. 19.1(c). This approach had the disadvantage of relying too heavily on intuition and so lacked the methodic reliability of the mass balance approach. Statistical techniques frequently do not work explicity in terms of time as a variable, which is inconvenient since engineers usually wish to make predictions as a function of time. Since both are useful when used properly, the polymer reactor engineer should be able to use both the statistical and deterministic methods.

——Laurence R L, et al. Mathematical Modelling of Polymerization Kinetics.
in: McGreavy C, ed. Polymer Reactor Engineering.
New York: Blackie Academic & Professional, 1994. 87-89

Words and Expressions

sequence distribution			（聚合物结构）序列分布
rational	['ræʃənl]	a.	合理的
moment	['məumənt]	n.	矩
interpretation	[intə:pri'teiʃən]	n.	解释
discrete	[dis'kri:t]	a.	离散的，分离的
postulation	[pɔstju'leiʃən]	n.	假定
via	['vaiə]	prep.	经过，通过
avenue	['ævə,nju]	n.	方法，途径
glean	[gli:n]	v.	收集
tedious	['ti:djəs]	a.	单调乏味的
colligative	['kɔligeitiv]	a.	综合的，依数性的
truncate	['trʌŋkeit]	v.	截短［断］
suffice	[sə'fais]	v.	足够，满足需要
bypass	['baipɑ:s]	n.; v.	旁路；绕过
probabilistic	[,prɔbəbi'listik]	a.	概率的
predominant	[pri'dɔminənt]	a.	主要的，有影响的
intuition	[,intju:'iʃən]	n.	直觉
explicity	[iks'plisiti]	n.	清楚，明确
recursive	[ri'kə:siv]	a.	回归的，递归的

Phrases

in some sense　某种意义上　　　　　　　　　　|　inasmuch as　因为

Notes

① "For a particular reactor type this mechanism may lead via the mass balance equations to a set governing the evolution of the concentration of polymer with a given chain length."译为："对于特定的反应器类型，这种聚合机理应用质量守恒理论到反应动力学方程组，这些方程决定了给定链长聚合物浓度的变化规律。"

② "Inasmuch as colligative methods of analysis yield data related to the moments, it would appear that the solution of a truncated set of moment equations might suffice to give an adequate description of a polymerization."译为："因为分析综合方法得到了有关矩的结果，从而有可能求解简化的矩方程组给出适当的聚合反应过程描述。"

③ "Often these techniques require extensive use of computers or the best they can offer is moments of the distribution."译为："这些方法需要广泛使用计算机，其最大好处是能得到分布的矩。"

④ "Cases arise when the polymerization can easily be thought of as some kind of classical statistical process such as Markov or Poisson process."译为："当把聚合反应简单地看作某种古典的统计过程，如马尔可夫链或泊松过程，出现了概率统计聚合机理研究方法。"

Exercises

1. *Complete the following according to the text*
   (1) Which mathematical techniques are most useful in modelling polymerization reactors? (i) ＿＿＿＿＿＿＿
   ＿＿＿＿＿＿＿＿＿＿＿＿＿＿＿＿＿＿(ii)＿＿＿＿＿＿＿＿＿＿＿＿＿＿＿＿＿＿．
   (2) "This" in "This opens several avenues …" (paragraph 3) refers to ＿＿＿＿＿＿＿＿＿＿＿＿＿＿＿
   ＿＿＿＿＿＿＿＿＿＿＿＿＿＿＿＿＿＿＿．
   (3) The statistical methods had the disadvantages of (i) ＿＿＿＿＿＿＿＿＿＿＿＿＿＿＿＿＿＿＿
   ＿＿＿＿＿＿＿＿＿＿＿＿＿＿＿＿＿＿(ii)＿＿＿＿＿＿＿＿＿＿＿＿＿＿＿＿＿＿＿＿＿＿＿．
   (4) The article aims to ＿＿＿＿＿＿＿＿＿＿＿＿＿＿＿＿＿＿＿＿＿＿＿＿＿＿＿＿＿＿．

2. *Put the following words into English*
   分子量分布　　分布函数的矩　　共聚物组成和序列分布　　聚合物链长
   数学变换方法　马尔可夫链理论　概率统计方法　　　　　　质量守恒

3. *Put the following into Chinese*

   There are two alternative generic starting points for the employment of mathematical analysis in polymerization. One can start from the mass balance equations on the reactor of interest. This leads to differential or algebraic difference equations, depending on reactor type, which must be solved on the rate constants of model. This is a systematic approach, readily adapted to the analysis of different reactor types. This method produces an infinite set of mass balance equations, one for each possible polymer chain length, 1 to ∞. The solution of any particular set of such equations may be difficult and involved, but there is a catalog of solution methods available (Ray & Laurence, 1977).

   Alternatively, statistical methods can be useful. These rely on picturing the polymer chain as growing by a process of selecting monomer from the reaction mixture according to some statistical distributions. The success of this approach depends on a correct, and frequently intuitively based, correspondence of the polymerization process with a postulated stochastic process. It is therefore less easily adaptable to different reactor configurations since the appropriate correspondence may be difficult to find or, in fact, may not exist. On the other hand, when statistical methods do work, they usually lead to results in a comparatively simple way and can sometimes reveal features of a polymerization system that a mass balance approach does not. The fully equipped polymerization engineer should be able to take advantage of both mass balance and statistical techniques.

Reading Materials

Numerical or Direct Integration Techniques for Modelling Polymerization Kinetics

As already noted, a polymerization reaction is described by an infinite set of species mass balance equation. The difficulties encountered in the numerical integration of this set of rate equations are most apparent by looking at a simple example.

Anionic polymerization of styrene in a solvent, e. g. tetrahydrofuran (THF), with a butyl lithium initiator results in a chain-growth polymerization. The mechanism is very simple and therefore provides a very good system on which to illustrate various alternative solution techniques. The initiation reaction is very fast so that if I_0 moles of initiator are added at time $t = 0$, then for all practical purposes they are converted immediately to "living" polymers of chain length I, P_{10}, so the kinetics of initiation can be neglected. Furthermore, in a reaction mixture free from impurities, termination and transfer are negligible. The propagation reaction is the only one of interest, leading to a very simple mechanism:

$$P_j + M \xrightarrow{k_j} P_{j+1}, \quad j = 1, 2, 3, \cdots \qquad (19.1)$$

The rate constant k_j refers to the jth propagation step on a chain. The material balances for the various polymer species in a batch reactor then result in an infinite set of equations for the various species in the reactor

$$\frac{dM}{dt} = - \sum_{n=1}^{\infty} k_n P_n M \qquad (19.2)$$

$$\frac{dP_1}{dt} = -k_1 P_1 M \qquad (19.3)$$

$$\frac{dP_j}{dt} [k_{j-1} P_{j-1} k_j P_j] M \qquad (19.4)$$

with initial conditions

$$\left. \begin{array}{l} P_1 = I_0 \\ P_j = 0, \ j > 1 \\ M = M_0 \end{array} \right\} \text{at } t = 0 \qquad (19.5)$$

Since there are no termination reactions and all the growth chains are created at $t = 0$, it is obvious that the number of polymer molecules remains constant. This is verified by summing equations (19.3) and (19.4) for all j. The result is

$$\frac{d \sum_{n=1}^{\infty} P_n}{dt} = 0 \qquad (19.6)$$

So $\sum P_j = I_0$. The number of growing polymer molecules is equal to the number of initiator molecules added to the reaction mixture.

No real simplification is possible until an assumption is made. Fundamental to all analyses of polymerization kinetics is the assumption of equal reactivity, i.e. $k_j = k$ for all $j > 0$. This proves to be a reasonable assumption. The possibility that the rate constants are dependent on polymer chain length has been discussed by Zeman and Amundson (1965) and Saidel and Katz (1967). More recently, Kumar (1985) have analysed the effect of unequal reactivity in step-growth polymerizations.

Unequal reactivity, that is chain-length dependent reactivity, is often important in high conversion free radical polymerization. However, this need not be considered here.

A further simplification is a change in the time variable. Defining❶

$$\tau = \int_0^\infty k(t')M(t')dt'$$

leads to the following form of equation (19.4):

$$\frac{dP_j}{dt} = k[P_{j-1} - P_j] \qquad (19.7)$$

with $P_j(0) = 0$ for $j \geq 2; P_1(0) = I_0$

Thus, the goal is to examine various methods for extracting information on the chain length distribution from a set of kinetic equations in the form of equation (19.7).

As mentioned above, the obvious step is to solve equation (19.7) sequentially, i.e. start with P_1, then substitute it into the equation for P_2, solve it and so on. For the case of (19.7) this can be done analytically and it becomes clear there is a general form of the solution for P_j. In more complex cases this cannot be done analytically but can still be carried out numerically in the style of Liu and Amundson (1961). The simple system (19.7) is also amenable to a direct analytical solution via the Laplace transform, $\overline{P}_j$

$$\overline{P}_j(\lambda) = \int_0^\infty P_j(\tau)e^{-\lambda\tau}d\tau \qquad (19.8)$$

Applying the Laplace transform to equation (19.7) and the initial condition gives:

$$\lambda\overline{P}_j + \overline{P}_j = \overline{P}_{j-1} \qquad (19.9)$$

or

$$\overline{P}_j = \frac{1}{\lambda+1}\overline{P}_{j-1}$$

The general solution to this difference equation is

$$\overline{P}_j = K\left[\frac{1}{\lambda+1}\right]^{j-1} \qquad (19.10)$$

where K is a constant which can be determined from the normalization condition, that is the condition that in this case the total number of polymers is always constant at I_0. Therefore:

$$\sum_{n=1}^\infty P_j = I_0 \qquad (19.11)$$

---

❶ This change of time variables, first used in polymerization analysis by Dostal and Mark (1936), is known as the *eigenzeit* transformation. It allows for a much easier solution of equation (19.4) since the monomer concentration is a continuously changing quantity in a batch reactor. If the temperature control is poor, the reaction rate constant would also change.

and for the Laplace transform

$$\sum_{n=1}^{\infty} \overline{P}_j = \frac{I_0}{\lambda} \tag{19.12}$$

leading to

$$K \sum_{n=1}^{\infty} \left(\frac{1}{1+\lambda}\right)^j = \frac{K}{\lambda} = \frac{I_0}{\lambda} \tag{19.13}$$

so that $K = I_0$. Inserting this in (19.10) gives

$$\overline{P}_j = I_0 \left(\frac{1}{1+\lambda}\right)^j \tag{19.14}$$

and inverting the Laplace transform gives

$$P_j = I_0 \frac{\tau^{j-1} e^{-\tau}}{(j-1)!} \tag{19.15}$$

This is the Poisson distribution which has the properties discussed in more detail in some literature.

——Laurence R L, et al. Mathematical Modelling of Polymerization Kinetics. in: McGreavy C, ed. Polymer Reactor Engineering. New York: Blackie Academic & Professional, 1994. 94-96

Words and Expressions

butyl	['bjuːtil]	n.	丁基
lithium	['liθiəm]	n.	锂
mole	[məul]	n.	摩尔
termination	[təːmi'neiʃən]	n.	终止（反应）
amenable	[ə'miːnəbl]	a.	应服从的，有责任的
anionic polymerization			阴离子聚合
Laplace transform			拉氏变换（数学）
as already noted			如上所述
as mentioned above			如上所述
difference equation			差分方程

UNIT 20　Polymerization Process Instrumentation

One of the first considerations in establishing a strategy for controlling polymerization reactors is to categorize all system inputs and outputs into those which are to be controlled, those which may be adjusted to achieve this control, and those which are beyond the control of the designer. The cause-and-effect relationships in a polymerization reactor are depicted in Fig. 20.1.

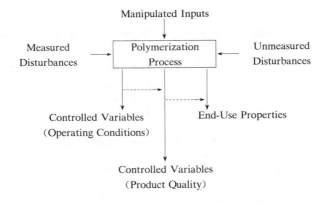

Fig. 20.1

Polymerization process variables

The system outputs may be divided into three categories: end-use properties, controlled variables affecting product quality, and controlled variables specifying operating conditions. So that the final product may meet the required specifications for a given application, the polymer must have certain end-use properties. End-use properties determine the suitability of a polymer for a specific application. These properties may be well defined (e. g. tensile strength), or they may be empirical measures of suitability in a given application (e. g. wear testing for traffic paints). End-use properties such as solubility, bulk density, extrudability, etc. , determine the saleability of the polymer. Polymer appearance factors such as color, refractive index, particle size and shape, are also important in some cases. In most cases, the end-use properties are not measurable on-line. Other outputs must be controlled so as the produce a polymer having the desired end-use properties.

From a control standpoint, the most important variables are those which ultimately affect the end-use properties. These will be referred to as controlled variables affecting product quality. The most important of these are MW, MWD, monomer conversion, copolymer composition distribution, copolymer sequence distribution, and degree of branching. Most of these variables are not measurable on-line. The common approach is to control those variables which are measurable, to estimate those which are estimable and control based on the estimates, and to fix those which cannot be estimated by controlling the inputs to the process. ① Close-loop control involves the adjustment of some manipulated variable (s) in response to a deviation of the associated control variable from its desired value. The purpose of closed-loop control is to bring the controlled variable to its desired value and maintain it at that point. Those variables which are not controllable in a closed-loop sense are maintained at their desired values (as measured by laboratory or other off-line measurement) by controlling all the identifiable input in order to maintain an unmeasured output at a constant value.

The next category of process outputs are controlled variables specifying operating conditions. Some examples of these are temperatures, pressures, and flow rates, associated with the process. These variables are most often measurable, and are closed-loop controlled.

The inputs to the polymerization system can be categorized as manipulated variables or disturbance variables. The manipulated variables are those which are adjusted, either automatically or manually, to maintain the controlled variables at their desired values. Common manipulated variables in polymerization processes include coolant or heating medium flow rates, gas or liquid flow rates for pressure control, feed rate of monomer, solvent, or initiator, and agitator speed.

The disturbance variables are those over which the control engineer has no control. ② Disturbances may be stochastic (random) or deterministic. Stochastic disturbances arise from the natural variability of the process. Examples are short-term variations in flow rates caused by mechanical inaccuracies. Deterministic disturbances arise from known causes, and they usually occur at longer intervals. Examples are lot-to-lot variations in feedstock quality③ and changes in production rates mandated by the operation of some upstream or downstream process. Although the cause of such disturbances may be known, the disturbances themselves cannot be eliminated because of constraints external to the system. Some disturbances, stochastic and deterministic, may be measurable, but by definition they cannot be eliminated. However, the effect of such disturbances on the final product can be eliminated by compensating for them by adjusting the manipulated variables. ④ This is the function of regulatory control.

The process variables do not always fall into such neat categories. For instance, temperature may be manipulated to adjust average molecular weight. In this case temperature is the manipulated variable for an MW control loop, but may at the same time be the controlled variable for a temperature control loop which uses the flow rate of a coolant as a manipulated variable. In this case the value of the manipulated variable for the MW control loop (temperature) is the desired value for the temperature control loop. This, of course, is the notion of cascade control.

Not all process variables are measurable on-line; most end-use properties are not. Some controlled variables affecting product quality are measurable or they can be estimated; many are not. Even when the technology to measure these variables on-line exists, the cost of such sensors may be prohibitive. On-line gel permeation chromatography for the determination of MWD is an example. Almost all controlled variables are known because they are either measured or set by a control system (or both). As discussed above, disturbances may be measured or unmeasured.

One final point about closed-loop process control. Economic considerations dictate that to derive optimum benefits, processes must invariably be operated in the vicinity of constraints. ⑤ A good control system must drive the process toward these constraints without actually violating them. In a polymerization reactor, the initiator feed rate may be manipulated to control monomer conversion or MW; however, at times when the heat of polymerization exceeds the heat transfer capacity of the kettle, the initiator feed rate must be constrained in the interest of thermal stability. In some instances, there may be constraints on the controlled variables as well. Identification of constraints for optimized operation is an important consideration in control system design. Operation in the vicinity of constraints poses problems because the process behavior in this region becomes increasingly nonlinear.

In many cases, the capability to control polymerizations is severely limited by the state-of-the-art in measurement instrumentation. In other cases, the dynamic response of the instruments dictates the design strategy for the process.

——Schork F J. Reactor Operation and Control.
in: McGreavy C, ed. Polymer Reactor Engineering.
New York: Blackie Academic & Professional, 1994. 167-169

Words and Expressions

strategy	['strætidʒi]	n.	策略
tensile strength			抗张强度
extrudability	[ekstru:də'biliti]	n.	挤出变形能力，挤塑性
saleability	[seilə'biliti]	n.	销售性
standpoint	['stænd,pɔint]	n.	立场，观点
deviation	[,di:vi'eiʃən]	n.	背离，偏差［离］
identifiable	[ai'denti,faiəbl]	a.	可确认的
stochastic	[stə'kæstik]	a.	随机的
deterministic	[di'tə:mi,nistik]	a.	确定性的
feed stock			原料
cascade control			串级控制
mandate	['mændeit]	v.	要求
constraint	[kən'streint]	v.	约束
prohibitive	[prə'hibitiv]	a.	价格过高的，昂贵的
gel permeation chromatography			凝胶渗透色谱 GPC
vicinity	[vi'siniti]	n.	接近
pose	[pəuz]	v.	引起
state-of-the-art			（技术）发展水平

Phrases

short-term 短期的	fall into 归入某一范围
beyond the control of 失控	

Notes

① "The common approach is to control those variables which are measurable, to estimate those which are estimable and control based on the estimates, and to fix those which cannot be estimated by controlling the inputs to the process." 译为："常用的方法是对可测量的变量进行控制，对可估计的变量进行估计并根据估计值进行控制，对于不能估计的变量通过控制输入变量维持其不变。"

② "The disturbance variables are those over which the control engineer has no control." 译为："干扰变量指那些控制工程师不能控制的参量"。"control over" 控制。

③ "lot-to-lot variations in feedstock quality" 译为:"原料批次之间的质量波动。"
④ "However, the effect of such disturbances on the final product can be eliminated by compensating for them by adjusting the manipulated variables." 译为:"然而,这种干扰对最终产品质量的影响可通过调节控制参量得到补偿。"
⑤ "Economic considerations dictate that to derive optimum benefits, processes must invariably be operated in the vicinity of constraints." 译为:"从经济角度要考虑追求最佳的利益,但聚合过程必须在约束点附近操作。"

Exercises

1. *Complete the following according to the text*
   (1) There are three categories in polymerization process output variables. (i) _____ _____ . (ii) _____ . (iii) _____ .
   (2) Term "closed-loop control" means _____ _____ .
   (3) Common manipulated variables in polymerization process include _____ _____ .
   (4) Examples of deterministic disturbance are (i) _____ _____ (ii) _____ .
   (5) What is the notion of cascade control? _____ .
   (6) The capability to control polymerization limited by (i) _____ _____ (ii) _____ .

2. *Put the following words into English*
   控制策略   失控   适用性   在线测量
   聚合物支化度   调节变量   干扰变量   随机干扰
   约束条件   凝胶渗透色谱

3. *Put the following into Chinese*
   Two major problems arise in composition analysis: obtaining a representative sample from the reaction mass for analysis, and dealing with the "noise" in the analysis. The nature of composition analysis requires that a side stream or discrete sample of the system be obtained prior to the analysis. The problem in some cases is complicated by other factors, including the need to preclude exposure of the sample to air, changing composition during the sampling process due to reaction, or the physical character of the reactants which make sampling difficult. The result in many cases is that more time and effort are expended in development of the sampling method than is required to develop the analysis. The second problem is the presence of a significant level of noise which is often found in composition analysis, whether the analysis is run on-line or off-line. This uncertainty in the data can be often handled by signal filtering, although the control engineer must consider the quality of the measurements in designing a control scheme.

Reading Materials

Control of Semibatch Polymerization

Control of a semibatch reactor involves driving the reaction from an initial state to a specified final state in some manner which is judged to be the "best" in terms of productivity or product quality. Nonlinear model predictive control may be employed to drive the reaction along a trajectory which maximizes some predetermined objective function. Such a procedure requires an accurate model of the process, which is often not available. To deal with this lack of knowledge about the process, Schork and Houston (1987) have used a linear time series model which is updated on-line. Based on this model, a discrete algorithm is employed to calculate the optimal controls to be applied to the polymerizer. These controls are optimal for the current control interval only; no attempt is made to optimize the entire trajectory. This algorithm has been applied, using simulation, to the free-radical solution polymerization of methyl methacrylate (MMA) in a semibatch reactor. The reaction is carried out in a semibatch configuration. Monomer and solvent are charged to the reactor. A free-radical initiator is added during the course of the polymerization. The controller seeks, at each point, to maintain the number-average molecular weight at a desired value, while bringing the reaction to a specified monomer conversion in minimum time, by manipulating initiator feedrate and coolant jacket temperature. On-line determination of monomer conversion is available through densitometry or heat balance estimation. On-line determination of MWD is commercially available via the on-line gel permeation chromatography. Molecular weight information between discrete chromatograms may be supplied by open-loop prediction or by an extended Kalman Filter operating on the current time series model with rapid conversion feedback.

The baseline case against which this control scheme will be judged is the batch polymerization with PID temperature control. An initiator concentration of 0.0218 g·mol/L was chosen for the baseline because this is the amount which results in the correct average molecular weight. The batch time, molecular weight, polydispersity, and initiator used for this case are given in Table 20.1. The control scheme outlined above was applied to the system under the same conditions as the baseline case except that the initial initiator concentration was reduced to 0.0075 g·mol/L. The control objective was to regulate average molecular weight, then increase the rate of conversion (reducing polymerization time) when possible. The results are shown in Table 20.1. This control results in a 2% decrease in reactor batch time and a 55% decrease in total initiator use. The results show tight control of molecular weight distribution. The controller adds more initiator at the end of the batch to offset the gel effect. Since the average molecular weight produced is nearly constant, the resulting polydispersity is reduced to 1.626 from the baseline of 1.684.

One major advantage of adaptive predictive control is its ability to adapt to changing process conditions. To demonstrate this advantage, a process disturbance was applied to both the baseline and adaptive predictive control cases with all of the control parameters remaining the same. The disturbance selected was reduction in the initiator efficiency by 20%. This may be thought to represent the normal lot-to-lot variation in monomer and initiator reactivity. When this disturbance is applied to the baseline case, the batch time is increased by 10% and the molecular weight is increased by 9% (Table 20.1). When the same disturbance is applied to the adaptive predictive controller, more initiator is added to the reactor to compensate for the reduced initiator efficiency. Experiment results show that after an initial period of adjustment, the molecular weight remains on target throughout the batch cycle. Although 35% more initiator has been added to the reactor, there is little change in the molecular weight or polydispersity, as shown in Table 20.1.

Another process disturbance of considerable industrial concern is the level of residual inhibitor in the monomer. In order to keep MMA from polymerizing in storage, an inhibitor, such as hydroquinone, is added to the monomer in small quantities. Just before use, the inhibitor is removed from the monomer. Problems in the removal process may result in variations in the level of inhibitor remaining in the monomer charged to the reactor. This will result in variations in the polymerization rate. As before this disturbance was applied to both the baseline and adaptive predictive control case. When applied to the baseline control, this disturbance results in a 13% increase in batch time, a 7% increase in average molecular weight, and an increase in polydispersity from 1.684 to 1.708. The same disturbance applied to the adaptive predictive control case results in a 4% increase in batch time, but the molecular weight and polydispersity remained nearly the same (Table 20.1).

Table 20.1 Comparison of baseline and adaptive predictive control

Project	No disturbance		reduced initiator		inhibited monomer	
	baseline control	adaptive predictive control	baseline control	adaptive predictive control	baseline control	adaptive predictive control
batch time /h	2.44	2.38	2.72	2.34	2.76	2.47
molecular weight ($M_n/M_{n\ desired}$)	0.999	0.989	1.085	0.991	1.071	1.015
polydispersity	1.684	1.626	1.685	1.627	1.708	1.631
initiator added during polymerization /(g·mol/L)	0	0.0048	0	0.0065	0	0.0069
total initiator used /(g·mol/L)	0.0218	0.0099	0.0218	0.0108	0.0218	0.0110

Results indicate following improvement over isothermal batch polymerization: reduced initiator consumption, reduced batch time, good control of average molecular weight, and reduced product polydispersity. The noted improvements are even more significant when variations in initiator and/or monomer reactivity are considered. These improvements result from the presumption of high-quality on-line sensors for polymer properties (monomer conversion and number-average molecular weight), and an optimization-based controller which allows control objectives rather than just setpoints to be included.

——Schork F J. Reactor Operation and Control.
in: McGreavy C, ed. Polymer Reactor Engineering.
New York: Blackie Academic & Professional, 1994. 184-188

Words and Expressions

trajectory	['trædʒəktəri]	n.	轨迹，路径
predetermined	['priːdi'təːmind]	a.	预定的
algorithm	['ælgəriðəm]	n.	算法
simulation	[simju'leiʃən]	n.	模拟，仿真
densitometry	[densi'tɔmitri]	n.	密度法
baseline	['beislain]	n.	基线，原始资料
inhibitor	[in'hibitə]	n.	阻聚剂
presumption	[pri'zʌmpʃən]	n.	假定

UNIT 21 Reactor Scale-up

Scale-up is defined here as a presentation of the principles of scaling up and scaling down mixing systems. Pilot planting, on the other hand, is a determination of specific experiments and data analysis so that the controlling factors in the process can be uncovered. By knowing the controlling factors, it is possible to use the proper scale-up technique. Naturally, if more than one factor is involved, a consideration of all scale-up parameters must be carried out. In contrast to these two concepts, data from a Demonstration Plant usually yields only one point, and does not usually contain any information as to what would happen above or below the mixing conditions actually studied.

The first approach to scale-up was to use geometric similarity and dimensionless groups. There are 3 types of similarity-geometric, dynamic and kinematic. By working with dynamic similarity there are four groups of forces that are important shown in Table 21.1. These are the inertia force from the mixer and the fluid forces of viscosity, surface tension and gravity. It is impossible to keep the ratio of each of the individual fluid forces constant in scale-up with the same liquid, which is normally a pilot plant requirement. Therefore, we must pick and choose the ones that are important.

Table 21.1 Hydraulic similitude and force ratios

dimensionless groups relating force ratios	hydraulic similitude for model(M) and prototype (P)
$\dfrac{F_i}{F_v} = Re = \dfrac{ND^2\rho}{\mu}$	geometric: $\dfrac{X_M}{X_P} = X_R$
$\dfrac{F_i}{F_g} = F_r = \dfrac{N^2 D}{g}$	
$\dfrac{F_i}{F_\sigma} = W_e = \dfrac{N^3 d^2 \rho}{\sigma}$	dynamic: $\dfrac{(F_I)_M}{(F_I)_P} = \dfrac{(F_v)_M}{(F_v)_P} = \dfrac{(F_g)_M}{(F_g)_P} = \dfrac{(F_\sigma)_M}{(F_\sigma)_P} = F_R$

There have been some outstanding examples of scale-up solutions by means of these dimensionless groups. For example, in the power consumption of mixers the ratio of the inertia force of the mixer divided by the acceleration of the fluid correlates very well with the Reynolds number (which is the ratio inertia force in the mixer to viscous forces in the fluid). In heat transfer, the correlation Nusselt[①] as a function of the Reynolds number is another outstanding example. However, the dimensionless process group normally has a process result divided by some kind of system conductivity or some quantity related to the ease of carrying out the process result.[②] It is not possible to write dimensionless groups around very many mixing applications. This technique has limited usefulness in everyday operation.

Another technique is to obtain one data point at least on some particular scale for the process under consideration. Then everything else can be rationed to that data point. Table

21. 2 indicates how some of the ratios involved in mixing systems change on scale-up. As can be seen, it is normally not possible with geometric similarity to maintain the ratio of all the individual mixing parameters constant in scale-up. In fact, geometric similarity is nothing more than similarity of geometry, and does not ensure the similarity of any other fluid property in the system. [3] It is normally necessary to get some idea of which factors are controlling so that proper scale-up estimation can be made. Some examples have been cited preciously, particularly in the case of fermentation scale-up and the case of a gas-liquid-solid reaction.

Table 21. 2 Properties of a fluid mixer on scale-up

property	pilot scale 20 Gallons			plant scale 2500 Gallons	
P	1.0	125	3125	25	0.2
P/Vol	1.0	1.0	25	0.2	0.0016
N	1.0	0.34	1.0	0.2	0.04
D	1.0	5.0	5.0	5.0	5.0
Q	1.0	42.5	125	25	5.0
Q/Vol	1.0	0.34	1.0	0.2	0.04
ND	1.0	1.7	5.0	1.0	0.2
Re	1.0	8.5	25.0	5.0	1.0

One characteristic of gas-liquid scale-up is the fact that the linear superficial gas velocity normally increases. This means that the gas energy given up to liquid increases and can change the ratio of mixer horsepower to gas horsepower significantly.

Another thing that happens is the change in the shear rate relationship on scale-up. Fig. 21. 1 shows that as mixers are scaled up with geometric similarity, the maximum shear rate around the impeller goes up, while the average shear rate around the impeller goes down. This means that the shear rate distribution on big tanks is quite different from that be on small tanks, and must be compensated for in predicting full scale performance. There are other shear rate effects that are important and they must also be considered as scale-up, such as average tank shear rate and minimum tank shear rate.

——Nagata S. Mixing: Principle and Application. Tokyo: Kodansha Ltd. , 1975. 444-447

Fig. 21. 1

Curve illustrating that maximum impeller shear rate goes up while average impeller shear rate goes down on scale-up

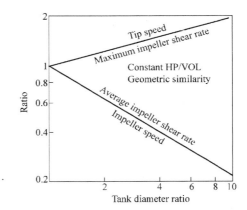

Words and Expressions

pilot planting			中试
kinematic	[ˌkainiˈmætik]	a.	运动学的
inertia	[iˈnəːʃiə]	n.	惯性
dimensionless group			无量纲数群
surface tension			表面张力
hydraulic	[haiˈdrɔːlik]	a.	水力的
similitude	[siˈmili,tjuːd]	n.	相似
fermentation	[fəːmenˌteiʃən]	n.	发酵
superficial	[ˌsjuːpəˈfiʃəl]	a.	表面的
full scale			工业规模的,全面的

Phrases

in contrast to [with] ... 和…对比,比较起来
give up to... 把…献给…,把…用来做…
compensate for... 补偿,弥补

Notes

① " Nusselt " 努塞尔数, $Nu = \dfrac{\alpha L}{\lambda}$ 其中 α 为传热系数,λ 为热导率,L 特性长度。

② "However, the dimensionless process group normally has a process result divided by some kind of system conductivity or some quantity related to the ease of carrying out the process result." 译为:"无论如何,无因次数群分析法通常描述成过程结果除以某种系统传递特性或有关过程结果简化处理的某些特征量。"

即:$\dfrac{过程结果}{系统的传递性质} = f\left(\dfrac{作用的力}{阻力}\right)$。

例如传热:$\dfrac{结果}{系统导热性} = f\left(\dfrac{作用力}{阻力}\right)$,$\dfrac{\alpha L}{\lambda} = \left(\dfrac{ND^2\rho}{\mu}\right)^x \left(\dfrac{C_P\mu}{\lambda}\right)^y \left(\dfrac{D}{d}\right)^z$。

③ "In fact, geometric similarity is nothing more than similarity of geometry, and does not insure the similarity of any other fluid property in the system." 译为:"事实上,几何相似仅仅是(大小反应器)几何形状相似而已,并不能保证系统中任何其他流动性质的相似。"

Exercises

1. *Complete the following sentences according to the text*

   (1) The fundamental purpose of a pilot plant is _____.

   (2) By working with dynamic similarity there are four groups of forces. (ⅰ) _____. (ⅱ) _____. (ⅲ) _____. (ⅳ) _____.

   (3) According Table 21.2, if a reactor scale up 125 times with geometric similarity and constant power/volume, The ratio of tip speed (ND) is _____.

   (4) The average shear rate on big tanks is _____ than small tanks scaling up with geometric similarity and constant power/volume.

2. *Put the following into English*

   雷诺准数　　　　中试工厂　　　　惯性力　　　　黏性力
   运动相似　　　　放大和缩小　　　搅拌桨叶端速度　剪切速率
   无量纲数群　　　工业规模反应器

3. *Put the following into Chinese*

   Scale up is defined as: the successful startup and operation of a commercial size unit whose design and operating procedure are in part based upon experimentation and demonstration at a smaller scale of operation.

The concept of successful must include production of the product at planned rate, at the projected manufacturing cost, and to the desired quality standards. Implicit in the term cost are not only the obvious factors such as the purchase prices for raw materials, the product yield, and the return on capital, but also the overall safety of the contemplated operation to plan personnel, the public, and the environment. The timing of project completion is also in most instances a critical factor. An experience in scale up that results in the startup being completed later than planed is not a very successful experience.

To be successful at the scale up of chemical process requires the utilization of a board spectrum of technical skills and a mature understanding of the total problem under study. Scale up procedures do not involve only technical decisions and compromises. The selected compromise always has an economic aspect since it is never possible to establish exactly what an industrial process should be. There are always restrictions of time and money availablity for the total development program of which scale up is only a part. Therefore, calculated risks will have to be taken in the design, construction, and startup of a "first commercial unit". The scope of the risks (and, the resulting financial uncertainties) will have to be considered against the additional expenses required to improve still further one's knowledge of the process.

Reading Materials

Succeed at Scale-up

Scale-up problems in industrial reactor mixing can be costly, but are all too common. Here's a proven procedure that avoids them.

Failure to properly scale-up mixing in batch and continuously stirred vessels remains a persistent problem in the process industries. Numerous causes of difficulty exist, but a pattern is evident: A lack of understanding of the process undermines many efforts. Mixing problems are seldom recognized, in part because they are not well understood, but also because they are not examined as quantitatively as other unit operations.

Correcting errors in scaling mixing operations is costly and sometimes impossible. Errors can cause losses of productivity, quality, and profit, and they also can lead to safety problems such as reactive-chemical incidents. Although technical problems cause most failures, nontechnical reasons also contribute to difficulties.

Scale-up of mixing can be easy or complex. Most difficulties arise when potential problems have not been well thought out. Hardest of all are multiphase processes in which the chemistry depends on condition of the phases.

Good quantitative tools and measures of performance are required to solve such problems. (See the accompanying article by Tatterson for an examination of these needs). My 25 years of experience with industrial mixing and scale-up problems has led me to identify some general troublespots and to develop a procedure for successful scale-up.

First of all, the engineer must clearly understand the role of the mixing. Is rapid attainment of uniformity critical to process success? The following are typical chemical processes that depend on rapid attainment of uniformity:

1. chemical reactors / polymerizers in which reaction kinetics are equal to or faster than the rate of mixing;

2. competing chemical reactions where poor mixing affects yields;

3. crystallizers that depend on uniform mixing to promote the growth of large uniform crystals;

4. reactions dependent on mass transport, such as coalescing and redispersing of liquid-liquid and gas-liquid mixtures.

For such processes, desired results can be achieved more easily in small equipment than in large equipment.

Contrast those with applications that are less sensitive to the needs of uniform mixing. These include:

1. heat transfer;
2. blending of miscible fluids;
3. reactors involved with slow chemical reactions;
4. suspension of solids.

These four can be considered noncritical applications, i.e., they can usually be scaled up with few difficulties.

Many engineers are most familiar with the latter, noncritical applications. Unfortunately, they get the impression that mixing is simple and can be treated casually. Thus, it is not surprising that so many scale-up problems occur.

Avoiding problems

For successful scale-up of mixing in industrial processes, a designer should follow six distinct steps:

1. define the process need;
2. identify all of the operational parameters;
3. review the process history;
4. select the important process parameters;
5. choose an initial equipment design—vessel design, impellers, impeller location, baffles, and points of feed and exit streams; and test the design relative to the process needs and assumptions and then fine tune it to meet the needs of the most important variables.

Many scale-up failures can be traced directly to the omission of one or more of these six steps.

——Leng D E. *Chemical Engineering Progress*. 1991, 6: 23

Words and Expressions

persistent	[pə'sistənt]	a.	持久稳固的
undermine	[ˌʌndə'main]	v.	破坏
nontechnical	[nɔn'teknikəl]	a.	非技术上的
troublespot	['trʌblspɔt]	n.	故障点
attainment	[ə'teinmənt]	n.	达到
crystallizer	['kristəˌlaizə]	n.	结晶槽
redisperse	[ˌriːdis'pəːs]	v.	再分散
blending	[blendiŋ]	n.	混合，共混，混炼
noncritical	[nɔn'kritikəl]	a.	非苛刻的
unfortunately	[ʌn'fɔːtʃənitli]	adv.	不幸地
impression	[im'preʃən]	n.	印象
distinct	[dis'tiŋkt]	a.	清楚的，独特的
omission	[əu'miʃən]	n.	省略
mass transport			质量传递，传质
think out			慎重考虑

UNIT 22 UNIPOL Process for Polyethylene

1. Process description

In recent years, the UNIPOL process① has become a popular commercial technology for linear polyethylene production (Burdett, 1988). In this process, the copolymerization of ethylene and α-olefins is carried out in a fluidized-bed reactor using a heterogeneous Ziegler-Natta or supported metal oxide catalyst. A schematic diagram of the reactor system is shown in Fig. 22.1. The feed to the reactor comprises ethylene, a comonomer (1-butene or a higher alpha-olefin), hydrogen and nitrogen. These gases provide the fluidization and heat transfer media and supply reactants for the growing polymer particles. The catalyst and a cocatalyst are fed continuously to the reactor. The fluidized particles disengage from the reactant gas in the expanded top section of the reactor. The unreacted gases are combined with fresh feed streams and recycled to the base of the reactor. Since the reaction is highly exothermic, heat must be removed from the recycle gas before it is returned to the reactor. The rate of polymer production is determined from an on-line heat balance. The mass of material in the bed is also calculated on-line using bed level and pressure measurements.② The conversion per pass through the bed is very low, making the recycle stream much larger than the fresh feed streams. Because polymer particles in the fluidized bed are mixed well and the conversion per pass③ is low, gas composition and temperature are essentially uniform throughout the bed. Periodically, the product discharge valve near the base of the reactor opens and the fluidized product flows into a surge tank. The unreacted gas is recovered from the product that proceeds downstream for further processing and distribution.④

Fig. 22.1

Gas-phase polyethylene reactor system

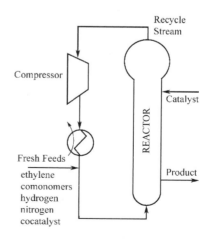

The melt index and density of the polymer in the bed depend on catalyst properties, reactant gas composition, and reactor temperature. The reactor is instrumented well with temperature, pressure, and flow sensors. Gas compositions are measured by on-line gas chromatographs. Melt index and density are measured every several hours in the quality control laboratory. These analyses require up to one hour. When the lab results become available, they are used to adjust the reactor operating conditions to ensure that on-specification polymer[5] is produced.

2. Models for melt index and density

Any scheme to predict melt index and density between measurements requires a model describing how these variables are affected by reactor operating conditions. If the reactor is operated near one set of operating conditions to produce a limited number of polymer products, then an empirical linear plant model will often suffice. However, one of the advantages of the UNIPOL process over traditional liquid-phase systems is the wider range of products that can be produced[6] (Burdett, 1988). The models developed for this application must be valid over the range of products made in the reactor. Thus, linear empirical models are not suitable.

A kinetic model describing molecular weight and copolymer composition development and their relationships to melt index and density is presented by Mc Auley et al (1990). While this model can predict MI, ρ, and production rate in an industrial reactor, the structure of this model (22 differential equations) is prohibitively complex for use in an on-line quality inference scheme. The approach taken in this article is to simplify the theoretical model so that it becomes appropriate for on-line use. Although several different comonomers are used to produce linear polyethylene in UNIPOL systems, it is uncommon to operate with ethylene and more than two comonomers in the reactor simultaneously. Hence, the simple models for MI and ρ are developed for ethylene, butene and one higher alpha-olefin (HAO) comonomer. Extensions to more comonomers are straightforward.

Unmeasured impurities and unmodeled disturbances can result in sustained offset between model predictions and measured quality variables. If such drifts in product quality are not accounted for in the control scheme, then large quantities of off-grade polymer can be produced. One way to alleviate this problem is to force the model to track the process by updating parameters and predictions recursively on-line. If the common sources of the expected mismatch are known, then this information can be used to choose which parameters remain constant and which are likely to change due to the disturbances. Theoretically-based models have an advantage over empirical models in that the designer may have some prior knowledge about which parameters require on-line updating. Usually only a few meaningful parameters need to be updated, thereby making the on-line schemes easier to maintain and monitor.

——McAuley K B, MacGregor J F. *AIChE Journal*. 1991, 37 (6): 825

Words and Expressions

α-olefins	['ælfə-'əulifin]	n.	α-烯烃
heterogeneous	[,hetərəu'dʒi:njəs]	a.	不同种类的，非均相的
Ziegler-Natta		n.	Z-N 催化剂
support	[sə'pɔ:t]	v.	（催化剂）负载化
schematic	[ski'mætik]	a.	示意性的
comonomer	[kɔ'mɔnəmə]	n.	共聚单体
butene	['bju:ti:n]	n.	丁烯
disengage	[,disen'geidʒ]	v.	脱离
surge tank			缓冲槽，聚料仓

surge	[səːdʒ]	n.; v.	汹涌，波动
melt index			融熔指数
instrument	['instrumənt]	v.	用仪器装备
scheme	[skiːm]	n.; v.	安排，计划
empirical	[em'pirikəl]	a.	经验的
differential	[,difə'renʃəl]	a.	微分的
inference	['infərəns]	n.	推论
straightforward	[,streit'fɔːwəd]	a.	简单的，直截了当的
sustain	[səs'tein]	v.	持续不变，维持
off-grade		n.	等外品
offset	['ɔfset]	n.; v.	偏移
alleviate	[ə'liːvieit]	v.	减轻
mismatch	[mis'mætʃ]	v.	不匹配
meaningful	['miːniŋful]	a.	意味深长的

Phrases

be not accounted for... 不引起对…的重视，不对 | …进行考虑

Notes

① "the UNIPOL process" 译为："联碳公司（UCC）开发的气相流化床聚乙烯工艺。"
② "The mass of material in the bed is also calculated on-line using bed level and pressure measurements." 译为："通过床层高度和压力测量也可在线计算反应器内的持固量。"
③ "the conversion per pass" 单程转化率。
④ "The unreacted gas is recovered from the product that proceeds downstream for further processing and distribution." 译为："未反应气体从产品中回收，产品进入下一工段进行进一步处理。"
⑤ "on-specification polymer" 特定规格的聚合物。
⑥ "However, one of the advantages of the UNIPOL process over traditional liquid-phase systems is the wider range of products that can be produced." 译为："然而，UNIPOL 工艺优于传统液相系统的特点之一是能生产更宽范围的聚合物产品。"（注：UNIPOL 工艺能生产全密度聚乙烯）

Exercises

1. *Complete the following according to the text*
   (1) The melt index and density of the polymer in the UNIPOL process depend on (ⅰ) _____ . (ⅱ) _____ . (ⅲ) _____ .
   (2) Linear empirical models for melt index and density are not suitable in UNIPOL system. Why? _____ .
   (3) One way to alleviate offset between model predictions and measurements is _____ .
   (4) Theoretically-based models have an advantage over empirical models in that _____ .
2. *Put the following into English*

   非均相催化剂   负载化催化剂   流化床反应器   线性低密度聚乙烯
   熔体流动速率   聚烯烃   单程转化率   出料阀
   在线气相色谱   经验与理论模型 长链 α-烯烃

3. *Put the following into Chinese*

   In the UNIPOL gas-phase process, reactor capacity is determined, to a large extent, by the heat removal capabilities of the circulating gas. A large gas stream is circulated through the fluidized bed where polymer is produced and discharged out of the reactor for downstream processing. The gas stream is then recompressed and sent through a cycle gas cooler to remove the heat of reaction. Before being sent back to the reactor, reactants are injected into the recycle gas stream. Removing the heat of reaction is the key to capacity and is highly dependent on the cycle gas composition as the different reactants, such as ethylene, butene, hexene and hydrogen, have much different physical properties.

These physical property differences, driven mainly by catalyst chemistry, diminish reactor capacity for hexene film resins, making them more expensive than their butene counterparts.

A patent was granted in the mid-1980's showing that it was possible to operate very stably with liquid in the inlet reactor gas stream. The heat removal capability was improved by taking advantage of the condensed liquid's latent heat of vaporization. This was very important for the economic production of hexene copolymers. As referenced by Ken Sinclair in his February Society of Plastics Engineers presentation, this same patent taught that the stable operating limit for condensed mode was around 10wt% liquid. Above this, the risk of hot spots, lump formation and reactor instability increase.

Reading Materials

EUROPEAN PATENT APPLICATION

(21) Application number: 90105621.8	(51) Int. Cl.^{5}. C08F 12/04 [1]
(22) Date of filing: 24.03.90	
(30) Priority: 30.03.89 JP 76496/89 (43) Date of publication of application: 03.10.90 Bulletin 90/40	(71) Applicant: IDEMITSU PETROCHEMICAL CO. LTD. 1-1, Marunouch1 3-chcme Chlyodaku Tokyo 100 (JP)
(84) Designated Contracting States: [2] AT BE CH DE ES FR GB IT	(72) Inventor: Yamamoto, Koj1 Idemltsu Petrochemlcal Co., Ltd. 1-1 Aneaklkalgan Ichihara-shl, Chiba-ken (JP) Inventor: Teshima, Hideo Idemitsu Petrochemical Co., Ltd. 1-1 Anesaklkalgan Ichihara-shl, Chiba-kon (JP)
	(74) Representative: Turk, Gllle, Hrabal Brucknerstrasse 20 D-4000 Dusseldorf 13 (DE)

(54) Process for producing a styrene-based polymer.

(57) Disclosed[3] is a process for producing a styrene-based polymer having a high syndiotactic configuration, which comprises polymerizing a styrene-based monomer with addition of catalyst in such amounts that the conversion at 30 minutes after the polymerization starts becomes 2% to 50%, and then continuing the polymerization with additions of the catalyst when the conversion exceeds 10%.

According to the process, the productivity can be improved while easily attaining safe running of the reactor and successful of controlling of reaction heat.

BACKGROUND OF THE INVENTION

1. Field of the Invention

The present invention relates to a process for producing a styrene-based polymer, for an efficient production of a styrene-based polymer having a stereostructure in which the chains of polymers are in a high syndiotactic configuration.

2. Description of the Related Arts

Styrene-based polymers having a stereostructure which is in atactic or isotactic configuration have heretofore been well known, but recently styrene-based polymers having a stereostructure of a high syndiotactic configuration have been developed one of which is disclosed in Japanese Patent Application Laid-Open No. 187708/1987. [4]

The reaction system, in which styrene-based polymers having syndiotactic configuration are under production, solidifies when the polymerization reaction proceeds to a conversion of approximately 20%, while the reaction continues further until a higher conversion is accomplished. By applying an appropriate shearing force at said solidification stage, polymers in a favorable powder form can be obtained. If the reaction rate at an early stage is too high, however. A large shearing force is required for inhibiting the formation of macro-particles, which is involved in a fear that the inside of the reactor might be wholly covered with solid polymers. There is also a fear of causing a melt fusion of polymers, since it is difficult to control the reaction heat caused by reaction at a high rate.

On the contrary if the reaction rate is lowered, for the purpose of inhibiting the formation of macro-particles, it takes much time to complete the polymerization and the production efficiency is lowered.

SUMMARY OF THE INVENTION

An object of the present invention is to provide a process for an efficient production of styrene-based polymers having syndiotactic configuration without adhesion of the polymers to the reactor or solidification of the polymers into cakes. Another object of the present invention is to provide a process for an efficient production of styrene-based polymers having syndiotactic configuration, by stable running of polymerization with low power consumption.

The present invention provides a process for producing a styrene-based polymer having a high syndiotactic configuration, which comprises polymerizing styrene-based monomers with addition of catalysts in such amounts that the conversion at 30 minutes after the polymerization starts becomes 2% to 50%, and subsequently containing the polymerization with addition of the catalysts when the conversion exceeds 10%.

COMPARATIVE EXAMPLE

Into a tank-type reactor having a capacity of 10 L a diameter of 200 mm, provided with an agitation blade of multipaddle type with a blade length of 190 mm, a paddle width of 25 mm, a blade angle of 30° (to the axis line), five-paddle blades, and an anchor-type paddle as the lowest paddle, a length of this axis line of 290 mm, 230 mm, 170 mm, 108 mm from the bottom, and the ends of paddles excepting the lowest two paddles being provided with scrapers having a length of 60 mm (the upper-most scraper), 72 mm (the second-tier scraper), 85 mm (the third-tier scraper), a width of 13 mm, a clearance of 2 mm from the inside wall of tank. 4 L of styrene as the starting material, 40 mmol of triisobutylaluminum, 40 mmol of methylaluminoxane, 0.2 mmol of pentamethylcyclopentadienyltrimethoxytitanium[5] as catalyst were placed, and reacted at 75 °C for two hours with the agitating blade running at 450 r/min. After the reaction was completed, the conversion of styrene was 30%.

Subsequently, a further 40 mmol of triisobutylaluminum, 40 mmol of methylaluminoxane, 0.2 mmol of pentamethylcyclopentadienyltrimethoxytitanium were added, and subjected to reaction for further three hours, and then the conversion of styrene was found to be 70%. The resulting styrene polymer (polystyrene) was in a favorable powder form having an average particle diameter of 0.3 mm. The weight average molecular weight of said styrene polymer was 623000, and the syndiotacticity in terms of racemic pentad determined by ^{13}C-NMR was 98%. ⑥

Claims

(1) A process for producing a styrene-based polymer having a high syndiotactic configuration, which comprises polymerizing a styrene-based monomer with addition of catalyst in such amounts that the conversion at 30 minutes after the polymerization starts becomes 2% to 50%, and then continuing the polymerization with additions of the catalyst when the conversion exceeds 10%.

(2) The process according to Claim 1 wherein the catalyst comprises a titanium compound and the reaction product of trialkylaluminum and water.

(3) The process according to Claim 2 wherein the reaction product is alkylaluminoxane.

(4) The process according to Claim 1 wherein the addition of the catalyst at the second time or thereafter is effected at the stage where the conversion is in the range of 15% to 90%.

(5) The process according to Claim 1 wherein the polymerization temperature is 150℃ or lower.

Words and Expressions

priority	[prai'ɔriti]	n.	优先权
representative	[,repri'zentətiv]	n.	代理人
syndiotactic	[sindaiə'tæktik]	a.	间同立构的
heretofore	[,hiətu'fɔ:]	a.	直到此时
stereostructure	[,stiəriə'strʌktʃə]	n.	立体结构
triisobutylaluminum	[tri'aisəu'bju:tilə'lju:miniəm]	n.	三异丁基铝
methylaluminoxane	['meθələ'lju:mən'ɔksənə]	n.	甲基铝氧烷
racemic	[rə'si:mik]	a.	外消旋的
claim	[kleim]	n.	(专利)保护权项

Notes

① "(51) Int. Cl.^{5}" 即 International patent classification 5th，国际专利分类法第五版。
　前面的数字 (51) 系专利说明书内容国际标准代码，如 (21) 指申请号。
② "Designated Contracting States" 欧洲专利的指定保护国，后续的国别用 ISO 代码表示。
③ "Disclosed is …" 译为："本发明…"专利说明书常用语。
④ "Japanese Patent Application Laid-Open" 一种日本专利。
⑤ "pentamethyl cyclopenta dienyl trimethoxytitanium" 译为："五甲基环戊二烯基三甲氧基钛。"
⑥ "^{13}C-NMR"
　^{13}Carbon Nuclear Magnetic Resonance Spectroscopy 碳-13 核磁共振，可用于聚合物结构的研究。

PART C

Processing, Properties and Applications of Polymer Material

UNIT 23　Polymer Processing

Polymer processing, in its most general context, involves the transformation of a solid (sometimes liquid) polymeric resin, which is in a random form (e. g. powder, pellets, beads), to a solid plastics product of specified shape, dimensions, and properties. This is achieved by means of a transformation process: extrusion, molding, calendering, coating, thermoforming, etc. The process, in order to achieve the above objective, usually involves the following operations: solid transport, compression, heating, melting, mixing, shaping, cooling, solidification, and finishing. Obviously, these operations do not necessarily occur in sequence, and many of them take place simultaneously.

Shaping is required in order to impart to the material the desired geometry and dimensions. It involves combinations of viscoelastic deformations and heat transfer, which are generally associated with solidification of the product from the melt. ①

Shaping includes: (1) two-dimensional operations, e. g. die forming, calendering and coating, and (2) three-dimensional molding and forming operations. Two-dimensional processes are either of the continuous, steady state type (e. g. film and sheet extrusion, wire coating, paper and sheet coating, calendering, fiber spinning, pipe and profile extrusion, etc.) or intermittent as in the case of extrusions associated with intermittent extrusion blow moulding. Generally, moulding operations are intermittent, and, thus, they tend to involve unsteady state conditions. Thermoforming, vacuum forming, and similar processes may be considered as secondary shaping operations, since they usually involve the reshaping of an already shaped form. In some cases, like blow molding, the process involves primary shaping (parison formation) and secondary shaping (parison inflation).

Shaping operations involve simultaneous or staggered fluid flow and heat transfer. In two-dimensional processes, solidification usually follows the shaping process, whereas solidification and shaping tend to take place simultaneously inside the mold in three dimensional processes. Flow regimes, depending on the nature of the material, the equipment, and the processing conditions, usually involve combinations of shear, extensional, and squeezing flows in conjunction with enclosed (contained) or free surface flows. ②

The thermo-mechanical history experienced by the polymer during flow and solidification results in the development of microstructure (morphology, crystallinity, and orientation distributions) in the manufactured article. The ultimate properties of the article are closely related to the microstructure. Therefore, the control of the process and product quality must be based on an understanding of the interactions between resin properties, equipment design, operating conditions, thermo-mechanical history, microstructure, and ultimate product properties. Mathematical modeling and computer simulation have been employed to obtain an understanding of these interactions. Such an approach has gained more importance in view of the expanding utilization of computer aided design/computer assisted manufacturing/computer aided engineering (CAD/CAM/CAE) systems in conjunction with plastics processing.

The following discussion will highlight some of the basic concepts involved in plastics shaping operations. It will emphasize recent developments relating to the analysis and simulation of some important commercial processes, with due consideration to elucidation of both thermo-mechanical history and microstructure development. More extensive reviews of the subject can be found in standard references on the topic (1~6).

As mentioned above, shaping operations involve combinations of fluid flow and heat transfer, with phase change, of a visco-elastic polymer melt. Both steady and unsteady state processes are encountered. A scientific analysis of operations of this type requires solving the relevant equations of continuity, motion, and energy (i.e. conservation equations).

——Austarita G, Nicolas L. Polymer prscessing and properties
New York: Plenum press 1984, 1-3

Words and Expressions

processing	['prəusesiŋ]	n.	加工,成型
dimension	[d(a)i'menʃən]	n.	尺寸
extrusion	[eks'tru:ʒən]	n.	挤出
mo(u)lding	['məuldiŋ]	n.	模塑
calendering	['kælindəriŋ]	n.	压延
coating	['kəutiŋ]	n.	涂覆,涂布
thermoforming	['θə:məufɔ:miŋ]	n.	热成型
shaping	['ʃeipiŋ]	n.	成型
viscoelastic	[viskəui'læstik]	a.	黏弹性的
heat transfer			热传递
die forming	['daifɔ:miŋ]	n.	口模成型
intermittent	[intə'mitənt]	a.	间歇式的
secondary shaping operation			二次成型（操作）
parison	['pærisən]	n.	型坯
squeeze	[skwi:z]		挤压

Phrases

in its most general context　在其最一般的情况下　　with due consideration to ...　适当考虑
in view of　由于,鉴于,考虑到

Notes

① "It involves combinations of viscoelastic deformations, ...from melt." 译为:"成型过程包括黏弹性形变和热传递,这种黏弹形变和热传递是和产品从熔体的固化（或冷却）相联系的。" "solidification" 对热塑性材料一般翻译成冷却,对热固性材料来讲译为固化。该句中 "which" 引出的是一个非限定性定语

从句，用以修饰"combinations of viscoelastic deformations and heat transfer."
② "Flow regime, depending on the nature of the material, ... or free surface flows."译为："根据材料的性质、设备和加工条件的不同，流动（形式）以及根据流动面的自由与否，通常包含剪切流动、拉伸流动及挤压流动。"（国内教材上将流动形式一般只分为剪切与拉伸流动）

Exercises

1. *Complete the following according to the text*

Polymer processing is a science of _____ of polymeric resin to _____ of specific shape. It involves development of new machines and processes. In order to convert a polymer to a desired article, the processes, generally speaking, involve these operations: solid transport, _____, _____, shaping, _____, etc. Obviously, these operation don't nesessarily occur in this order, many of them take place simultaneously.

Polymer processing is subdivided into auxiliary, main (primary) and secondary processes. Main processing operations are extrusion, _____, _____, _____, etc. Mixing, milling, pelletizing and drying are examples of auxiliary processes. Secondary processing involves the reshaping of an already shaped form. _____ and _____ are the typical examples of secondary processing.

Polymer processing is a complex process. It involves the melting of resins, melt flow, viscoelastic deformation, heat transfer and the formation of microstructure. Microstructures of the article _____ the processing conditions, and the end properties of the article are closely related to the microstructure. Therefore, the controls of the quality of the article and the selection of the processing conditions must be based on the understanding of _____ between resin properties, equipment design, operation conditions, microstructure and ultimate product properties.

2. *Translate the following paragraph into Chinese*

The methods used to fabricate polymer products have a strong influence on the properties and performance of the products. Of special concern are mechanical and optical properties. The variation in properties is largely associated with the difference in structural order in these products, the most important influence probably being that of molecular orientation. However, other physical phenomena may also have a significant effect, such as the thermodynamic state (including crystallinity and physical aging of glassy plastics) and the development of residual thermal stresses by rapid cooling. Also, optical characteristics are determined by structure within the polymer which has dimensions approximately equal to the wavelength of visible light, and by defect on surface.

3. *Put the following into English*

固体输送	热传递	产品质量
加工	成型	二次成型
黏弹形变	微观结构	相互作用

Reading Materials

Polymer Processing: Historical Survey

The beginning of plastics processing and extrusion processes is associated with the introduction of gutta-percha into England during the 1840s and its commercial development as insulation for electrical wire. One of the early pioneers of the new industry was Thomas Hancock's younger brother, Charles Hancock, one of the founders of the Gutta Percha Company. In his patents of 1846~1847, Charles Hancock described fabrication of gutta-percha using a processing technology similar to that developed in the rubber industry largely by his brother. He used a "pickle"-type masticator for compounding gutta-percha with additives including sulfur and softeners. He also sheeted with rollers and vulcanized the products with sulfur.

The first foamed plastics and rubber products were developed in 1846 in separate patents by the Hancock brothers. Charles Hancock (English Patent No. 11032) foamed gutta-percha using ammonium carbonate and similar compounds. William Brockedon and Thomas Hancock (English patent No. 11455) produced foamed products using sulfur chloride dissolved in a rubber or gutta-percha solution.

The first ram extrusion devices were described in the patents of 1845 by Richard A Brooman (English Patent No. 10582) and Henry Bewley (English patent No. 10825), which discussed the manufacture of gutta-percha thread, tubes and hose. Brooman's patent uses a five-hole die that produces five simultaneous continuous filaments which are extruded into a bath and taken up on a roll. Bewley's patent extruded tubes and hose. Charles Hancock, who was a partner of Bewley in the Gutta Percha Company, is said to have developed insulation coating for wire using Bewley's extrusion methods. Methods of coating wires are described in patents by Barlow and Forster (English Patent No. 12136) and by Siemens (English Patent No. 13062) in 1848~1850. The first great successes of gutta-percha were its application to electrical insulation of the Dover-Calais and trans-Atlantic cables.

The development of continuous extrusion of plastics using screw extruders began with gutta-percha and natural rubber and dates from the 1870s. The concept of screw pumping seems to be attributable to Archimedes. The earlier use of screw pumps in the soap industry is described in the patent literature. The first patent for screw extrusion is that of Matthew Gray of London in 1879 (English Patent No. 5056). Interestingly, the reason for the invention as cited by Gray is the existence of defects in coatings placed on wires. The extruder was fed from a two-roll mill or calendering device. There seems to have been independent developments of the screw extruder in Germany and the USA at about the same time, but Gray's patent is the first clear statement.

The next stage in the development of polymer processing methods came with the commercial development of cellulose nitrate as a plastics. The first moves in this direction during the 1860s by Alexander Parkes and Daniel Spill in England met with only limited success. Cellulose nitrate could not be melted and they used a range of volatile solvents that evaporated from their products. These left high levels of residual stresses which caused shrinking and cracking. Parkes and Spill had rubber-processing backgrounds and apparently used rubber-processing machinery. In the USA, John Wesley Hyatt and his brother Isaiah Smith Hyatt found that compounds or solutions of cellulose nitrate in nonvolatile camphor produced more desirable products. This was called Celluloid. The Celluloid Manufacturing Company was formed in the 1870s in Newark, New Jersey, to exploit this product and proved to be a great success. The Hyatts and their associates developed many important industrial processing operations to exploit Celluloid.

An 1872 patent by the Hyatt brothers (US Patent No. 133229) contains both the reinvention of the ram extruder and the first ram injection molding machine. They called this a stuffing machine. John Wesley Hyatt later described the use of complex multiple-cavity molds to be used in conjunction with the stuffing machine. This would either mold objects or coat cores of objects in the mold.

In an 1878 patent, John Wesley Hyatt (US Patent No. 204228) described the extrusion of Celluloid from the stuffing machine over a mandrel coated with a lubricant. This mandrel could be programmable and expand to produce complex hollow shapes. This led to the development of blow molding in 1881 by the Hyatts' colleague, William B Carpenter (US Patent No. 237168). Here, a preformed extruded tube is placed in a mold and is then expanded to fill the mold by pumping a heated fluid into the tube. These inventions were largely empoyled to produce a range of products including components of dolls and liners for pipes.

The 1880s saw the development of the synthetic fiber industry. Brooman's 1845 patent for the formation of gutta-percha thread sets out clear procedures for producing fibers from the melt. The synthetic fibers sold commercially in this period were produced from cellulose nitrate which could not be melted. A method of producing fibers by extruding acetic acid solutions of cellulose nitrate into a water or alcohol coagulation bath was described by Joseph Wilson Swan (English Patent No. 5978) in 1883. Swan's patent describes the later carbonization of the fibers with heat and thus represents the beginning of the carbon-fiber industry. Swan's application was filaments for incandescent lights. Shortly thereafter in France, the Count de Chardonnet (US Patent No. 394559) described a process for forming fibers from ether alcohol solutions by extruding the solutions into a water coagulation bath. De Chardonnet produced much finer fibers than Swan, he formed a company and commercialized them as an artificial silk. Later, de Chardonnet (US Patent No. 531158) described a dry spinning process in which the filaments were extruded into the air where the solvent was evaporated. Also, during the 1890s, using the system of and collaborating with Cross, Bevan and Beadle, Stearn invented a reactive spinning method in which cellulose is dissolved in a mixture of sodium hydroxide and carbon disulfide to form cellulose xanthate, which is extruded into an acid coagulating bath that regenerates the cellulose. This material became known as rayon.

The first truly synthetic plastics, phenol formaldehyde resins, were developed commercially by Leo Hendrik Baekeland, a Belgian immigrant to the USA, just before 1910. These were poured as low or intermediate molecular weight liquids into molds where they were polymerized into three-dimensional networks. Bakelite products were compression molded.

Large internal mixers were introduced into the rubber industry in the second decade of the twentieth century by Fernley H Banbury, an English immigrant to the UA (US Patent No. 1200070). These were made necessary by increased production needs, the toxic accelerators used to hasten vulcanization and the increasing use of large quantities of finely divided particles, especially in the new pneumatic-tire industry. To some extent, these represented a new generation of Hancock's "pickle" mixers suitable for the modern period of rubber processing.

By 1920, the essential features of the processing of rubber, plastics and fibers had been established. The years since then have seen additional major innovations. These are associated with the development of new synthetic thermoplastics, which were mostly introduced from 1920 to 1960. Screw extrusion devices using two co-or counter-rotating screws were developed and applied as mixing devices. Reciprocating screws were introduced and replaced rams in injection molding machines.

Numerous innovations in forming technology have been developed. In the late 1930s, Du Pont, notably through the efforts of Carothers, developed a commercial process for continuously forming highly oriented synthetic fibers from aliphatic polyamides (US Patent No. 2130948). This included the melt spinning and drawing processes.

Since the late 1940s, films have been produced by extruding melt from an annular die and expanding the melt with air. The basic patent of tubular film extrusion seems to be that of Fuller of the Visking Corporation issued in 1949 (US Patent No. 2461975).

Early process technologies mixed, pumped and shaped products. New process technologies have been developed since the mid-1960s, which go beyond this to produce products with desirable mechanical properties inducing certain types of molecular orientation. The manufacture of melt-spun and drawn fibers is a forerunner of these technologies. Slit film extrusion, notably of poly (ethylene terephthalate), followed by the subsequent continuous development of biaxial orientation on a tentering frame was developed by Du Pont investigators in the mid-1950s (US Patent No. 2728941). A generation later, Wyeth and Roseveare of Du Pont (US Patent No. 3849530) devised stretch blow molding of bottles where an inner mandrel vertically deforms as well as inflates a parison preform to produce a biaxially oriented bottle.

Most spectacular has been the invention of liquid crystalline process technology for aromatic polyamides by Kwolek, Morgan and their coworkers at Du Pont which has been applied to the manufacture of super high modulus and tensile strength fibers (US Patent No. 3671542).

Other developments have been the invention of processes for the formation of large foamed parts manufacture of multilayer film and sheet, and skincore (sandwich) injection-molded parts.

—— White J L. Encyclopedia of Material Science and Engineering. Vol 5. In: Bever M B, ed. Oxford: Pergamon Press, 1986. 3694

Words and Expressions

pickle	[pikl]	n.	空投鱼雷
masticator	['mæstikeitə]	n.	捏合机，素炼机
additive	['æditiv]	n.	添加剂
sulfur (sulphur)	['sʌlfə]	n.; vt.	硫黄，硫化
softener	['sɔfənə]	n.	软化剂
vulcanize	['vʌlkənaiz]	v.	硫化
ammonium carbonate			碳酸铵
cellulose nitrate			硝酸纤维素
residual stress			残余应力
stuffing machine			充填机
mandrel (mandril)	['mændrəl]	n.	型芯
coagulation bath			凝固浴
xanthate	['zænθeit]	n.	黄原酸酯（盐）
accelerator	[æk'seləreitə]	n.	促进剂

UNIT 24 Mechanical Properties of Polymers

The mechanical properties of polymers are of interest in all applications where polymers are used as structural materials. Mechanical behavior involves the deformation of a material under the influence of applied forces.

The most important and most characteristic mechanical properties are called moduli. A modulus is the ratio between the applied stress and the corresponding deformation. The reciprocals of the moduli are called compliances. The nature of the modulus depends on the nature of the deformation. The three most important elementary modes of deformation and the moduli (and compliances) derived from them are given in Table 24.1, where the definitions of the elastic parameters are also given. ① Other very important, but more complicated, deformations are bending and torsion. From the bending or flexural deformation the tensile modulus can be derived. The torsion is determined by the rigidity.

Table 24.1 Survey of elastic parameters and their definitions

elementary mode of deformation	elastic parameter	symbol	definition
isotropic (hydrostatic) compression	bulk modulus	K ①	hydrostatic pressure/volume change per unit volume $p/(\Delta V/V_n) = pV_n/\Delta V$
	bulk compliance or compressibility	k ($k=1/K$)	reciprocal of the foregoing
simple shear	shear modulus or rigidity	G	shear force per unit area/shear per unit distance between shearing surfaces = $(F/A)/\tan\gamma = \tau/\tan\gamma \approx \tau/\gamma$
	shear compliance	J ($J=1/G$)	reciprocal of the foregoing
uniaxial extension	tensile modulus or young's modulus	E	force per unit cross-sectional area/strain per unit length = $(F/A)/\ln(L/L_0) = \sigma/\varepsilon \approx (F/A)/(\Delta L/L_0)$
	tensile compliance	S ② ($S=1/E$)	reciprocal of the foregoing
any	poisson ratio	ν	change in width per unit width/change in length per unit length = lateral contraction/axial strain

① Often the symbol B is used.

② In older literature the symbol D is often used.

Cross-linked elastomers are a special case. Due to the cross-links this polymer class shows hardly any flow behavior. The kinetic theory of rubber elasticity was developed by Kuhn, Guth, James, Mark, Flory, Gee and Treloar. It leads, for Young's modulus at low strains, to the following equation:

$$E = 3RT\rho/M_{crl} = 3z_{crl}RT/V = 3C_0 \qquad (24.1)$$

where R——gas constant = 8.314 J/(mol·K);

M_{crl}——number average molecular weight of the polymer segments between cross-links;
ρ——density;
z_{crl}——average number of cross-links per structural unit $= M/M_{crl}$;
V——molar volume of structural unit;
C_0—— $z_{crl}RT/V$.

The paragraphs above dealt with purely elastic deformations, i. e. deformations in which the strain was assumed to be a time-independent function of the stress. In reality, materials are never purely elastic: under certain circumstances they have nonelastic properties. This is especially true of polymers, which may show nonelastic deformation under circumstances in which metals may be regarded as purely elastic.② It is customary to use the expression viscoelastic deformations that are not purely elastic. Literally the term viscoelastic means the combinations of viscous and elastic properties. As the stress-strain relationship in viscous deformations is time-dependent, viscoelastic phenomena always involve the change of properties with time. Measurement of the response in deformation of a viscoelastic material to periodic forces, for instance during forced vibration, shows that stress and strain are not in phase: the strain lags behind the stress by a phase angle δ, the loss angle. So the moduli of the materials, the complex moduli, include the storage moduli which determine the amount of recoverable energy stored as elastic energy, and the loss moduli which determine the dissipation of energy as heat when the material is deformed.③

——Vankrevelen D W. Properties of Polymers. New York: Elsevier Scientific Publishing Co., 1990. 367

Words and Expressions

deformation	[ˌdiːfɔːˈmeiʃən]	n.	变形
compliance	[kəmˈplaiəns]	n.	柔量
isotropic	[ˌaisəuˈtrɔpik]	a.	各向同性的
hydrostatic	[ˌhaidrəuˈstætik]	a.	流体静力学的
uniaxial	[ˈjuːniˈæksiəl]	a.	单轴的
torsion	[ˈtɔːʃən]	n.	转矩
nonelastic	[ˈnɔniˈlæstik]	a.	非弹性的
rupture	[ˈrʌptʃə]	n.	断裂
orientation	[ˈɔːriənˈteiʃən]	n.	定向
crystalline	[ˈkristəlain]	a.	结晶的
elastic parameter			弹性指数
Young's modulus			杨氏模量

Phrases

under the influence of   在…的影响下
lags behind   滞后于
be subjected to   使受到
in reality   事实（实际）上

be a function of   是…的函数
set in   开始
have a pronounced effect on   对…有重大影响

Notes

① "The three most important elementary modes of deformation and the moduli (and compliances) derived from them are given in Table 24.1, where the definitions of the elastic parameters are also given." 译为：" 表 24.1 给出了三种最重要的基本变形方式和由这些变形推导得到的模量（或柔量），表中也给出了弹性指数的定义。" 句中 "where" 引导的非限制定语从句修饰 Table 24.1。

② "This is especially true of polymers, which may show nonelastic deformation under circumstances in which metals may be regarded as purely elastic." 译为:"这对于聚合物来说尤其如此,在金属被认为纯弹性的情况下聚合物可能表现出非弹性变形。"句中第一个"which"引导的是非限制定语从句,修饰"polymers";第二个"which"引导的是限制性定语从句,修饰"circumstances"。

③ "So the moduli of the materials, the complex moduli, include the storage moduli which determine the amount of recoverable energy stored as elastic energy, and the loss moduli which determine the dissipation of energy as heat when the material is deformed." 译为:"因此,材料的模量即复合模量包括储存模量和损失模量,其中储存模量决定了以弹性能量储存的可恢复能量的多少,损失模量决定了材料变形时以热量形式散失的能量。"句中插入的 the complex moduli 是前边 moduli 的同位语。

Exercises

1. *Link the English words with the corresponding Chinese meanings*

    deformation   各向同性的
    crystallite   流体静力学的
    torsion   变形
    isotropic   微晶
    rupture   结晶的
    hydrostatic   转矩
    crystalline   断裂

2. *Translate the following paragraph into Chinese*

    The strength properties of solids are most simply illustrated by the stress-strain diagram, which describes the behavior of a homogeneous specimen of uniform cross section subjected to uniaxial tension. If the material fails and ruptures at a certain tension and a certain small elongation, it is called brittle, and its theoretical ultimate strength is of the order of $\sigma_{th} \approx E/10$, where E is Young's modulus. If permanent or plastic deformation sets in after elastic deformation at some critical stress, the material is called ductile, and the yield stress is $\sigma_y = \sigma_{max} \sim E^n$, where $n \approx 0.75$.

Reading Materials

Mechanical Properties of Uniaxially Oriented Polymer

If an isotropic polymer is subjected to an imposed external stress, it undergoes a structural rearrangement called orientation. In amorphous polymers this is simply a rearrangement of the randomly coiled chain molecules (molecular orientation). In crystalline polymers the phenomenon is more complex: crystallites may be reoriented or even completely rearranged; oriented recrystallization may be induced by the stress applied. The rearrangements in the crystalline material may be read from the X-ray diffraction patterns.

Nearly all polymeric objects have some orientation: during the forming (shaping) of the specimen the molecules are oriented by viscous flow and part of this orientation is frozen in as the object cools down. But this kind of orientation is negligible compared with the stress-imposed orientation applied in drawing or stretching processes.

Orientation has a pronounced effect on the physical properties of polymers. Oriented polymers have varying properties in different directions, i.e. they are anisotropic.

Uniaxial orientation is accomplished by stretching a thread, strip or bar in one direction, usually this process is carried out at a temperature just above the glass transition point. The polymer chains tend to line up parallel to the direction of stretching, although in reality only a small fraction of the chain segments becomes perfectly oriented.

Uniaxial orientation is of the utmost importance in the production of man-made fibers since it provides the desired mechanical properties like modulus and strength. In addition, it is only by stretching or drawing that the spun filaments become dimensionally stable and lose their tendency to creep at least at room temperature. The filaments as spun possess a very low orientation unless spinning is performed at extreme velocities. Normally a separate drawing step is required to produce the orientation necessary for optimum physical properties. In practice a drawing machine consists of two sets of rolls, the second running faster depending on the stretch ratio, which is usually about four.

As mentioned already, the effects of orientation on the physical properties are considerable. They result in increased tensile strength and stiffness with increasing orientation. Of course, with increasing orientation the anisotropy of properties increases too. Oriented fibres are strong in the direction of their long axis, but relatively weak perpendicular to it.

If the orientation process in semi-crystalline fibres is carried out well below the melting point (T_m), the thread does not become thinner gradually, but rather suddenly, over a short distance: the neck. The so-called draw ratio is the ratio of the length of the drawn to that of the undrawn filament; it is about 4 to 5 for many polymers, but may be as high as 40 for linear polyolefins and as low as 2 in the case of regenerated cellulose.

The degree of crystallinity does not change much during drawing if one starts from a specimen with a developed crystallinity (before drawing); if on the other hand, the crystallinity of the undrawn filament is not, or only moderately, developed, crystallinity can be greatly induced by drawing. The so-called "cold drawing" (e. g. nylon 6.6 and 6) is carried out more or less adiabatically. The drawing energy involved is dissipated as heat, which causes a rise of temperature and a reduction of the viscosity. As the polymer thread reaches its yield-stress, it becomes mechanically unstable and a neck is formed.

During the drawing process the crystallites tend to break up into microlamellae and finally into still smaller units, possibly by unfolding or despiralizing of chains. Spherulites present tend to remain intact during the first stages of drawing and often elongate into ellipsoids. Rupture of the filament may occur at spherulite boundaries; therefore it is a disadvantage if the undrawn thread contains spherulites. After a first stage of reversible deformation of spherulites, a second phase may occur in which the spherulites are disrupted and separate helices of chains (in the case of polyamides) become permanently arranged parallel at the fibre axis. At extreme orientations the helices themselves are straightened.

——Vankrevelen D W. Properties of Polymers. New York: Elsevier Scientific Publishing Co., 1990. 423

Words and Expressions

anisotropic	[,ænisə'trɔpik]	a.	各向异性的
spin	[spin]	v.	纺
adiabatic	[,ædiə'bætik]	a.	绝热的
microlamellae	[,maikrə'læmilei]	n.	微薄片 (microlamella 的复数)
ellipsoid	[i'lipsɔid]	n.	椭圆体
helices	['helisi:z]	n.	螺旋 (helix 的复数)
spherulite	['sferjulait]	n.	球晶

UNIT 25　Thermal Properties of Polymers

The heat stability is closely related to the transition and decomposition temperature, i. e. to intrinsic properties. By heat stability is exclusively understood the stability (or retention) of properties (weight, strength, insulating capacity, etc.) under the influence of heat.① The melting point or the decomposition temperature invariably form the upper limit; the "use temperature" may be appreciably lower.

The way in which a polymer degrades under the influence of thermal energy in an inert atmosphere is determined, on the one hand, by the chemical structure of the polymer itself, on the other hand, by the presence of traces of unstable structures.

Thermal degradation does not occur until the temperature is so high that primary chemical bonds are separated. For many polymers thermal degradation is characterized by the breaking of the weakest bond and is consequently determined by a bond dissociation energy. Since the change in entropy is of the same order of magnitude in almost all dissociation reactions, it may be assumed that also the activation entropy will be approximately the same. This means that, in principle, the bond dissociation energy determines the phenomenon. So it may be expected that the temperature at which the same degree of conversion is reached will be virtually proportional to this bond dissociation energy.②

The process of thermal decomposition or pyrolysis is characterized by a number of experimental indices, such as the temperature of initial decomposition, the temperature of half decomposition, the temperature of the maximum rate of decomposition, and the average energy of activation. The heat resistance of a polymer may be characterized by its "initial" and "half" decomposition.

There are two types of thermal decomposition: chain depolymerization and random decomposition. The former is the successive release of monomer units from a chain end or at a weak link, which is essentially the reverse of chain polymerization;③ it is often called depropagation or unzipping. This depolymerization begins at the ceiling temperature. Random degradation occurs by chain rupture at random points along the chain, giving a disperse mixture of fragments which are usually large compared with the monomer unit. The two types of thermal degradation may occur separately or in combination; the latter case is rather normal. Chain depolymerization is often the dominant degradation process in vinyl polymers, whereas the degradation of condensation polymers is mainly due to random chain rupture.

The overall mechanism of thermal decomposition of polymers has been studied by Wolfs et al. The basic mechanism of pyrolysis is sketched in Fig. 25.1.

In the first stage of pyrolysis (<550℃) a disproportionation takes place. Part of the decomposing materials is enriched in hydrogen and evaporated as tar and primary gas, the rest forming the primary char. In the second phase (>550℃) the primary char is further decomposed, i. e. mainly dehydrogenated, forming the secondary gas and final char. During the disproportionation reaction, hydrogen atoms of the aliphatic parts of the structural units are "shifted" to "saturate" part of the aromatic radicals. The hydrogen shift during disproportionation is highly influenced by the nature of the structural groups.

——Vankrevelen D W. Properties of Polymers. New York: Elsevier Scientific Publishing Co., 1990.725; 641

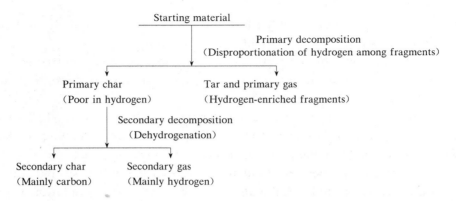

Fig. 25.1

Basic mechanism of pyrolysis

Words and Expressions

intrinsic	[in'trinsik]	a.	固有的
entropy	['entrəpi]	n.	熵
depropagation	[ˌdiːprɒpə'geiʃən]	n.	降解
unzippering	[ʌn'zipəriŋ]	n.	开链
depolymerization	[ˌdiːpɒliməraiˈzeiʃən]	n.	解聚
pyrolysis	[pai'rɒlisis]	n.	热解
char	[tʃɑː]	n.	炭
dehydrogenate	[diːˈhaidrədʒəneit]	v.	使脱氢
bond dissociation energy			键断裂能
random decomposition			无规降解

Phrases

at the ceiling temperature　在最高限度的温度下　　　be closely related to　与…密切相关
be proportional to　与…成比例　　　　　　　　　　at a weak link　在薄弱连接处
be enriched in　充满

Notes

① "By heat stability is exclusively understood the stability (or retention) of properties (weight, strength, insulating capacity, etc.) under the influence of heat" 译为："通过热稳定性可完全理解诸如重量、强度、绝缘能力等性能在热作用下的稳定性或保留率。" 这是一个含被动语态的倒装强调句，主语是 "the stability"。

② "So it may be expected that the temperature at which the same degree of conversion is reached will be virtually proportional to this bond dissociation energy." 译为："因此，可推出达到同一转化率对应的温度事实上与键断裂能成比例。" 句中 "which" 引导的定语从句修饰 "temperature"。

③ "The former is the successive release of monomer units from a chain end or at a weak link, which is essentially the reverse of chain polymerization."译为:"前者(链的解象)是单体单元在链端或薄弱连接处的连续释放过程,实质上是链式聚合的逆过程。"句中"which"引导的同位语从句修饰前边的"former"。

Exercises

1. *Put the following into English*

无规降解   链的解象   热解
解聚   脱氢   断裂能

2. *Fill the following brackets according to the pyrolysis mechanism described in the text*

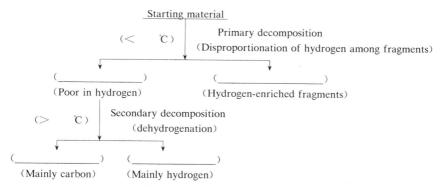

Reading Materials

Requirements for Heat Resistance

Heat resistance is the capacity of a material to retain useful properties for a stated period of time at elevated temperatures (≥230℃) under defined conditions, such as pressure or vacuum, mechanical load, radiation, and chemical or electrical influences at temperatures ranging from cryogenic to above 500℃. Both reversible and irreversible changes can occur. In a reversible change, for example, as a polymer under load approaches the glass-transition temperature T_g, deformation occurs without change in chemical structure. Reversible changes occur primarily as a function of T_g, which for the purposes of this article, i. e., for high temperature structural polymers, must be above 230℃. The maximum-use temperature for an amorphous or semicrystalline structural resin usually depends on T_g rather than the crystalline melt temperature T_m. A semicrystalline polymer can exhibit substantial loss of mechanical properties near the T_g, depending upon the degree of crystallinity. The T_m is usually so high that in its vicinity chemical degradation occurs. Irreversible changes alter the chemical structure. For example, exceeding the thermal stability results in bond breaking.

The chemical factors which influence heat resistance include primary bond strength, secondary or van der Waals bonding forces, hydrogen bonding, resonance stabilization, mechanism of bond cleavage, molecular symmetry (structure regularity), rigid intrachain structure, and cross-linking and branching. The physical factors include molecular weight and molecular weight distribution, close packing (crystallinity), molecular (dipolar) interactions, and purity.

The primary bond strength is the single most important influence contributing to heat resistance. The bond dissociation energy of a carbon-carbon single bond is ~350 kJ/mol

(83.6 kcal/mol), and that of a carbon-carbon double bond is ~610kJ/mol (145.8 kcal/mol). In aromatic systems, the latter is even higher. Known as resonance stabilization, this phenomenon adds 164~287kJ/mol (39.2~86.6 kcal/mol). As a result, aromatic and heterocyclic rings are widely used in thermally stable polymers.

Secondary or van der Waals bonding forces provide additional strength and thermal stability. Dipole-dipole interaction and H bonding contribute 25~41 kJ/mol (6.0~9.8 kcal/mol) toward molecular stability and affect the cohesion energy density, which influences the stiffness, T_g, melting point, and solubility. Thus, heat-resistant polymers often contain polar groups, e. g., —CO—, —SO$_2$—, that participate in strong intermolecular association. Polymers containing electron-withdrawing groups, e. g., —CO—, as connecting groups are generally more stable than those containing electron-donating groups, e. g., —O—.

The mechanism of bond cleavage also influences thermal stability. In polysiloxanes, for example, the energy of the silicon-oxygen single bond is ~445kJ/mol (106.4kcal/mol), and that of the silicon-carbon single bond ~328kJ/mol (78.4kcal/mol). Although the Si—C bond would be expected to cleave at high temperatures more readily than the Si—O bond, the latter breaks at high temperatures to form low molecular weight cyclic siloxanes because this degradation route is energetically favored.

Molecular symmetry or regularity of the chemical structure arises when moieties are joined in the same position in each repeat unit. The presence of isomers lowers T_g. Rigid intrachain structure refers to substitution of the aromatic or heterocyclic ring. Although para-oriented polymers have the highest thermal stability and T_g, they have the lowest solubility and processability.

Cross-linking improves heat resistance of a polymer primarily because more bonds must be cleaved in the same vicinity for the polymer to exhibit a weight loss or reduction in mechanical properties. Crystalline regions in a polymer serve as cross-links. Branching in a polymer tends to lower the thermal stability.

Molecular weight and molecular-weight distribution influence mechanical and physical properties. Higher molecular-weight polymers are more heat resistant than lower molecular-weight materials because of more entanglements and the ability to accommodate more chain cleavage without significant property reduction. Low molecular-weight polymers exhibit lower T_gs.

——Hergenrother P M. Encyclopedia of Polymer Science and Engineering. vol. 7. Editor-in Chlf Kroschwitz J I: 2nd ed. New York: John Wiley and Sons, 1985. 639

Words and Expressions

cryogenic	[,kraiə'dʒenik]	a.	低温的
reversible	[ri'və:səbl]	a.	可逆的
resonance	['rezənəns]	n.	共振
cleavage	['kli:vidʒ]	n.	断裂
intrachain	['intrə'tʃein]	a.	链内的
dipolar	[dai'pəulə]	a.	偶接的
heterocyclic	[,hetərə'saiklik]	a.	杂环的
association	[ə,səuʃi'eiʃən]	n.	缔合（作用）
polysiloxane	[,pɔli'siləksein]	n.	聚硅烷
moiety	['mɔiəti]	n.	部分，半个
isomer	['aisəumə]	n.	异构体
silicon	['silikən]	n.	硅

UNIT 26 Polymer Melts

1. Dynamics of polymer melts

It is now accepted that polymer chains are strongly intertwined and entangled in the melt; the dynamic behavior of such a system has been reviewed (*Graessley* 1974), but is only poorly understood. Thermodynamically, the chains are essentially ideal, as was first realized by Flory (1949). Their freedom of motion results from the presence of a *correlation hole* around each flow unit, within which the concentration of similar units from other chains is reduced. The presence of these correlation holes, and of the ideal but entangled nature of chains in the melt, has been confirmed by neutron-scattering experiments (*Cotton* 1974).

Methods commonly used for measuring the viscosity of polymer solutions and melts are listed in Table 26.1. The most important of these methods involve rotational and capillary devices (*Van Wazer* 1963, *Whorlow* 1979, *Dealy* 1981, 1983).

Table 26.1 Summary of methods for measuring viscosity

method	approximate useful viscosity range (Poise)	method	approximate useful viscosity range (Poise)
capillary pipette	$10^{-2} \sim 10^3$	falling coaxial cylinder	$10^5 \sim 10^{11}$
falling sphere	$1 \sim 10^5$	stress relaxation	$10^3 \sim 10^{10}$
capillary extrusion	$1 \sim 10^8$	rotating cylinder	$1 \sim 10^{12}$
parallel plate	$10^4 \sim 10^9$	tensile creep	$10^5 \sim >10^{12}$

2. Molecular weight and shear dependence

The most important structural variable determining the flow properties of polymers is molecular weight or, alternatively, chain length, Z (the number of atoms in the chain). Although early data (*Flory* 1940) suggested that $\lg \eta$ was proportional to $Z^{1/2}$, it is now well established (*Fox* 1956) for essentially all polymers studied that, for values of Z above a critical value Z_c,

$$\lg \eta = 3.4 \lg \overline{Z}_w + k \qquad (26.1)$$

where k is temperature dependent. This equation is valid only for shear stress sufficiently low ($10^2 \sim 10^3$ dynes/cm^2) that the viscosity is Newtonian.[①] The weight-average chain length $\overline{Z}_w$ is usually assumed to be the appropriate average for the above conditions. P. G. de Gennes (1979) discussed the unusual exponent 3.4 in terms of scaling concepts, considering it a major unsolved problem. Theories to explain it, including his own reptation model (*de Gennes* 1971), based on the wriggling motion of a chain inside a "tube" formed by its neighbors, fail to predict the exponent closer than 3.0 with no obvious reason for the discrepancy.[②]

For chain lengths below Z_c, which is about 600 for many polymers, the viscosity is found to depend upon a power of $\overline{Z}_w$ (and hence $\overline{M}_w$) in the range 1.75~2.0.[③] In this range, shear rate has little effect on viscosity.

While the Newtonian melt viscosity is determined by $\overline{M}_w$ as described in Eq 24.1, the dependence of viscosity on shear rate also depends upon the molecular-weight distribution. The drop in melt viscosity below its Newtonian value begins at a lower shear rate and continues over a broader range of shear rates for polymers with broader distributions of molecular weight. At sufficiently high shear rates, the melt viscosity appears to depend primarily on $\overline{M}_n$ rather than $\overline{M}_w$. Qualitative information, at least, about the molecular-weight distribution can be obtained from melt viscosity-shear rate studies.

3. Flow instabilities

At shear stresses in the neighborhood of 2×10^6 dynes/cm² for many polymers, instabilities in the flow appear, with the result that the upper Newtonian region is rarely realized in bulk polymers. These instabilites are manifested as a striking and abrupt change in the shape of the polymers stream emerging from the capillary of the rheometer. At and above a critical stress, the shape of the emerging stream changes from that of a regular cylinder to a rough or distorted one.

4. Temperature dependence of viscosity

For pure liquids, it was found many years ago that most of the change in viscosity with temperature is associated with the concurrent change in volume. This observation led to theories of viscosity based on the concept of free volume, whose application to polymers has been discussed in many other publications. A major result of these theories is the WLF equation, in which the temperature dependence of melt viscosity is expressed in terms of the glass transition temperature T_g (or another reference temperature) and universal constants. Since the terms describing the variation of melt viscosity with temperature and with molecular weight are independent, the WLF equation can be combined with Eq 26.2 to yield the relation, for low shear rates,

$$\lg \eta = 3.4 \lg \overline{Z}_w - \frac{17.44(T-T_g)}{51.6+T-T_g} + k' \qquad (26.2)$$

where k' is a constant depending only on polymer type. This equation holds over the temperature range from T_g to about $T_g + 100\ k$.

——Billmeyer F W. Jr. Textbook of Polymer Science. 3rd ed. New York: A Wiley-Interscience Publication, John Wiley & Sons, 1984. 303

Words and Expressions

intertwine	[ˌintəˈtwain]	v.	交缠，盘绕
entangle	[inˈtæŋgl]	v.	纠缠，缠绕
thermodynamically	[ˌθəːməudainæmikli]	adv.	热力学地，热动力地
correlation hole			相关孔隙
neutron-scattering			中子散射
variable	[ˈvɛəriəbl]	n.	变量
shear	[ʃiə]	v.	剪切
exponent	[eksˈpəunənt]	n.	指数，幂
scaling concept			标度概念
reptation model			蛇行模型
wriggle	[ˈrigl]	v.	蠕动，蜿蜒
discrepancy	[disˈkrepənsi]	n.	误差，偏差
rheometer	[riːˈɔmitə]	n.	流变仪
critical stress			临界应力
free volume			自由体积
constant	[ˈkɔnstənt]	n.	常数

Phrases

be proportional to 与…成比例	be associated with 与…有关系
be combined with 与…结合	with the result that 因而，从而

Notes

① "This equation is valid only for shear stress sufficiently low ($10^2 \sim 10^3$ dynes/cm^2) that the viscosity is Newtonian." 译为："只有当剪切应力足够小（$10^2 \sim 10^3$ dynes/cm^2）时，该方程式才适用，此时流体为牛顿流体。"

② "Theories to explain it, including... closer than 3.0 with no obvious reason for the discrepancy." 译为："解释这一现象的各种理论，包括 de Gennes 自己于1971年提出的'蛇行模型'（该模型基于'蚯蚓蠕动'，即一根分子链处于由它相邻分子所构成的'管道'中），由于没有明显的理由解释这一偏差，都不能预测比3.0更接近的指数。" "with no obvious reason for the discrepancy." 为"with"引导的原因状语。

③ "the viscosity is found to depend upon a power of $\overline{Z}_w$ (and hence $\overline{M}_w$) in the range $1.75 \sim 2.0$." 译为："发现黏度依赖于 $\overline{Z}_w$（因而也是 $\overline{M}_w$）的 $1.75 \sim 2.0$ 次幂。"

Exercises

1. Complete the following sentences according to the text
    (1) When melt, the polymer chains are strongly _____ and _____ .
    (2) The most important structural variable determining the flow properties of polymers is molecular weight or, alternatively, _____ _____ , Z (the number of atoms in the chain).
    (3) We can get the quantitative information about the _____ from melt viscosityshear rate studies.

2. Put the following words into English

缠结（$v.$）	中子散射	毛细管
指数	剪切速率	流变仪

Reading Materials

Thermodynamics of Melting Polymers

In polymers the crystalline melting point T_m is a phase change similar to that observed in low-molecular-weight organic compounds, metals, and ceramics.

The (Gibbs) free energy of melting is given by

$$\Delta G_m = \Delta H_m - T \Delta S_m \tag{26.3}$$

At the crystalline melting point T_m, $\Delta G_m = 0$; So

$$T_m = \frac{\Delta H_m}{\Delta S_m} \tag{26.4}$$

Now ΔH_m is the energy needed to overcome the crystalline bonding forces at constant temperature and pressure, and is essentially indepedet of chain length for high polymers. For a given mass or volume of polymer, however, the shorter the chains, the more "randomized" they become upon melting, giving a higher ΔS_m. Thus, the crystalline melting point decreases with decreasing chain length, and in a polydisperse polymer the distribution of chain lengths gives a distribution of melting points.

Equation (26.4) also indicates that chains that are strongly bound in the crystal lattice, that is, have a high ΔH_m, will have a high T_m, as expected. Also, the stiffer and less mobile chains, those that can "randomize" less upon melting and therefore have a low ΔS_m, will tend to have a higher T_m's.

Example: Discuss how the crystalline melting point varies with n in the "nylon n" series.

Increasing values of n dilute the nylon linkages that are responsible for interchain hydrogen bonding and thus lower the crystalline melting point. As n goes to infinity, the structure approaches that of linear polyethylene. This represents the asymptotic minimum T_m, with the chains held together only by van der Waal's forces.

Table 26.2 illustrates the variation in T_m and some other properties with n for some commercial members of the series.

Table 26.2 Variation of properties with n for nylon n's

n	T_m/℃	ρ/(g/cm³)	tensile strength/MPa	water absorption/% in 24 h
6	216	1.14	82.74	1.7
11	185	1.04	55.16	0.3
12	177	1.02	51.71	0.25
∞ (PE)	135	0.97	37.92	nil

——Rosen S. Fundamental Principles of Polymeric Materials. New York: A Wiley-Interscience Publication, 1971. 95

Words and Expressions

ceramics	[si'ræmiks]	n.	陶瓷
crystal lattice			晶格
randomize	['rændəmaiz]	v.	（使）无规分布
infinity	[in'finiti]	n.	无穷大
asymptotic	[æsimp'tɔtik]	a.	渐近的

UNIT 27　Processing and Fabrication of Thermoplastics

Processing and fabrication describe the conversion of materials from stock form (bar, rod, tube, pellet, sheet and so on) to a more or less complicated artefact. Polymer materials have been proved especially amenable to variety of extrusion and moulding techniques. In particular the following principal processing and fabricating operations for thermoplastics now enable products and components of complex shape to be mass-produced on a very large scale.

The most important processing and fabricating techniques for thermoplastics exploit their generally low melting temperatures and shape the materials from the melt. Extrusion and injection moulding are the most widely used processes. The screw extruder accepts raw thermoplastics material in pellet form and carries it through the extruder barrel; the material is heated by contact with the heated barrel surface and also by the mechanical action of the screw, and melts. The melt is compressed by the taper of the screw and is ultimately extruded through the shaped die to form tube, sheet, rod or perhaps an extrusion of more complicated profile. The screw extruder works continuously and the extruded product is taken off as it emerges from the die for reeling or cutting into lengths. It is important that the melt viscosity should be sufficiently high to prevent collapse or uncontrolled deformation of the extrudate when it leaves the die, and there may be water or air sprays at the outlet for rapid cooling. High melt viscosities can be obtained by using materials of high molar mass. The rate of cooling of an extrusion may determine the degree of crystallinity in a crystalline polymer, and hence affect mechanical and other properties.

Injection moulding describes a process in which polymer melt is forced into a mould, where it cools until solid. ① The mould then separates into two halves to allow the product to be ejected; subsequently the parts of the mould are clamped together once more, a further quantity of melted material is injected and the cycle repeated. The injection end of the machine is most commonly an Archimedean screw (similar to that of a screw extruder) which can produce, once per cycle, a shot of molten polymer of predetermined size and then inject it into the mould by means of a reciprocating ram action. Injection moulding provides a particularly effective way of obtaining complex shapes in large production runs.

As the size and/or aspect ratio of an injection moulding increase it becomes more difficult to ensure uniformity in the polymer during injection and to maintain a sufficient clamping force to keep the mould closed during filling. ② The reaction injection moulding process has been developed to overcome both these problems, essentially by carrying out most of the polymerizing reaction in the mould.

Blow moulding represents a development of extrusion in which hollow articles are fabricated by trapping a length of extruded tube (the parison) and inflating it within a mould. ③ The simple extruded parison may be replaced by an injection-moulded preform. Hollow articles including those of large dimensions may also be produced by rotational moulding (or rotocasting). A charge of solid polymer, usually powder, is introduced into a mould which is first heated to form a melt. The mould is then rotated about two axes to coat its interior surface to a uniform thickness.

Polymeric materials in continuous sheet form are often produced and subsequently reduced in thickness by passing between a series of heated rollers, an operation known as calendering. Thermoforming employs suction or air pressure to shape a thermoplastic sheet heated above its softening temperature to the contours of a male or female mould. Certain thermoplastics can be shaped without heating in a number of cold-forming operations such as stamping and forging commonly applied to metals.

All these methods of fabrication are essentially moulding processes. Cutting techniques (embracing all the conventional machining operations: turning, drilling, grinding, milling and planing) can also be applied to thermoplastics, but they are used much less widely. These techniques are most commonly applied to glassy polymers such as PMMA and to thermoplastic composites such as glass-filled PTFE, although softer materials such as PE and unfilled PTFE have excellent machinability.

Many thermoplastics may be satisfactorily cemented either to similar or dissimilar materials. A number of inert and insoluble polymer materials such as PTFE, PCTFE and some other fluoropolymers, PE, PP and some other polyolefins are amenable to cementing only after vigorous surface treatment. For the lower melting temperature thermoplastics welding provides an important alternative for joining parts of the same material, for example in fabricating pipework. The thermoplastics of higher melting temperature (which include PTFE of the common engineering polymers and also some of the specialised heat-tolerant materials such as the polyimides) cannot be satisfactorily fabricated by the principal extrusion and injection moulding processes. These materials are shaped by sintering powdered polymer in pressurised moulds, a process which causes the polymer particles to coalesce.

——Christopher Hall. Polymer Materials. London: The Macmillan Press Ltd., 1981. 162-165

Words and Expressions

fabrication	[ˌfæbrɪˈkeɪʃən]	n.	制造，二次加工
stock	[stɔk]	n.	原料，坯料
artefact (artifact)	[ˈɑːtɪfækt]	n.	人工制品
injection moulding			注塑
screw extruder			螺杆挤出机
barrel	[ˈbærəl]	n.	机筒，料筒
taper	[ˈteɪpə]	n.	锥体，锥度
die	[daɪ]	n.	塑模，口模
profile	[ˈprəʊfaɪl]	n.	外形，轮廓
collapse	[kəˈlæps]	vi.; n.	皱缩，塌瘪
extrudate	[eksˈtruːdɪt]	n.	挤出物
clamp	[klæmp]	vt.	夹（卡）紧，定位
Archimedean screw			阿基米德螺旋（杆）
shot	[ʃɔt]	n.	注（射）料量，物料量
reciprocating	[rɪˈsɪprəkeɪtɪŋ]	a.	往复的
ram	[ræm]	n.	柱塞，活塞
reaction injection moulding			反应注射成型
preform	[ˈpriːfɔːm]	n.	预成型坯，锭料
rotational moulding			旋转模塑
rotocasting	[ˌrəʊtəˈkɑːstɪŋ]	n.	离心浇铸
stamping	[ˈstæmpɪŋ]	n.	冲压成型，模压
forging	[ˈfɔːdʒɪŋ]	n.	锻造

turning	['tə:niŋ]	n.	车工工艺，车削
drilling	['driliŋ]	n.	钻（孔）
grinding	['graindiŋ]	n.	磨（光）
milling	['miliŋ]	n.	铣
planing	['pleiniŋ]	n.	刨（平）
PMMA		n.	聚甲基丙烯酸甲酯，有机玻璃
PTFE		n.	聚四氟乙烯
cement	[si'ment]	v.	粘接
PCTFE		n.	聚三氟氯乙烯
welding	['weldiŋ]	n.	焊接
pipework	['paipwə:k]	n.	管道工程
heat-tolerant		a.	耐热的
sintering	['sintəriŋ]	n.	烧结，热压结
coalesce	[ˌkəuə'les]	v.	凝聚（结）

Phrases

(be) amenable to　适合于…的
(to) take off …　拿开，移开
on a very large scale　很大规模地

Notes

① "Injection moulding describes a process in which polymer melt is forced into a mould, where it cools until solid." 句中 "in which…, where…solid." 是一个带介词的限定性定语从句，用来修饰 "process"。这个限定性的从句中又包含了一个非限定性的定语从句，即逗号后面的 "where it cools until solid"，这个非限定性的定语从句用来进一步说明 "mould"（模具）中的情况。译为："注塑描述的是以强力使聚合物熔体进入模具并在模具中冷却至固体的一种加工过程。"

② "As the size and/or aspect ratio of an injection moulding increase it becomes more difficult to ensure uniformity in the polymer during injection and to maintain a sufficient clamping force to keep the mould closed during filling." 此句中 "it" 引出的主句中两个不定式短语 "to ensure…and to maintain…" 是真正的主语，后一个不定式短语 "to maintain…" 中又包含了一个起目的状语作用的不定式短语 "to keep…"。此句可译为："当注塑（产品）的尺寸或纵横比增大时，很难确保注射时聚合物的均匀性和维持足够的闭合力以使模具在注料时保持闭合。"

③ "Blow moulding represents a development of extrusion in which hollow articles are fabricated by trapping a length of extruded tube (the parison) and inflating it within a mould." 此句中 "in which…" 引出的定语从句是用来定义主语 "Blow moulding" 的，是割裂式定语从句。因谓语比较简单，而定语从句比较长，把从句移到谓语后面，以保持句子匀称。此句可译为："吹塑是挤塑的一个发展，在吹塑成型中，中空的制品采用放入一段挤出的管坯并将其在模具中吹胀的方法制造出来。"

Exercises

1. *Complete the summary of the text. Choose no more than three words from the passage for each blank*

   Processing and fabrication describe the _____ of materials from stock form to artefact. The most important processing and fabricating techniques for thermoplastics exploit their generally _____ and shape them from melt. These techniques enable products and components of _____ to be mass-produced.

   Extrusion and injection moulding are the most widely used processes. Extrusion is accomplished by melting the material and forcing the melt through _____. Injection moulding describes a process in which polymer melt is forced into a _____, where it _____. The reaction injection moulding is a development of injection moulding.

Hollow articles are produced by _____ or rotational moulding. Continuous sheet materials are often produced in operation known as _____ . There are several other processing methods for thermoplastics such as thermoforming and cold-forming.

Cutting techniques can also be applied to some thermoplastics. Many thermoplastics are amenable to _____ . For lower melting thermoplastics _____ provides an important alternative for joining parts of the same material. The thermoplastics of higher melting temperature are shaped by _____ .

2. *Find out synonyms and line between them*

   blow   contour
   exploit   embrace
   include   stock
   parison   inflate
   profile   preform
   raw material   employ

3. *Put the following words into English*

   螺杆挤出机   长径比   型坯   口模
   反应注射模塑   压延   吹塑   粘接

4. *Put the following words into Chinese*

   collapse   rotocasting   heat-tolerant   clamp
   sintering   reciprocating   welding   coalesce

Reading Materials

Single-Screw Extrusion

The principal methods used to process thermoplastic materials into finished or semifinished products are, in order of importance: screw extrusion, injection moulding, blow moulding and calendering. The reason for the outstanding importance of extrusion is that almost all thermoplastics pass through an extruder at least once during their lives, if not during fabrication of the end product, then during the homogenization stage. Also, in most modern injection moulding machines, the preparation of melt for injection is carried out in what is essentially a screw extruder. Similarly, blow moulding and calendering are often post-extrusion operations.

The extrusion process is one of shaping a molten polymeric material by forcing it through a die. By far the most common method of generating the required pressure, and usually of melting the material as well, is by means of one or more screws rotating inside a heated barrel. While the form of the die determines the initial shape of the extrudate, its dimensions may be further modified—for example, by stretching—before final cooling and solidification takes place.

For a typical single-screw extrusion process, solid material in the form of either granules or powder is usually gravity fed through the hopper, although crammer-feeding devices are sometimes used to increase feed rates. After entering a helical screw channel via the feed pocket, the polymer passes in turn through the feed, compression and metering sections of the screw. The channel is relatively deep in the feed section, the main functions of which are to convey and compact the solids. The compression section owes its name to its progressively decreasing channel depth. Melting occurs there as a result of the supply of heat from the barrel and mechanical work from the rotation of the screw. The shallow metering section, which is of constant depth, is intended to control the output of the machine, generate the necessary delivery pressure and mix the melt. On leaving the screw, the melt is usually forced through a perforated breaker plate. A renewable screen pack, consisting of layers of metal gauze, may also be used; like the breaker plate, this serves to hold back unmelted polymer, metal particles or other foreign matter, and to even out temperature variations. Single-screw extruders are essentially screw viscosity pumps, which rely on material adhering to the barrel for their conveying and pumping action. Thanks to high melt viscosities, pressures of 40 MN/m^2 or more can be achieved at the delivery end of the screw.

The screw is held in position by an axial thrust bearing and driven by an electric motor via a reduction gearbox. Screw speeds are generally within the range 50 to 150 rev/min, and it is usually possible to vary the speed of a particular machine over at least part of this range. The radial clearance between the flights of the screw and the barrel is small, and the surfaces of both are hardened to reduce wear.

Barrel and die temperatures are maintained by externally mounted heaters, typically of the electrical-resistance type. Individual heaters or groups of heaters are controlled independently via thermocouples sensing the metal temperatures, and different zones of the barrel and die are often controlled at different temperatures. The region of the barrel around the feed pocket is usually water cooled to prevent fusion of the polymer feedstock before it enters the screw channels. Cooling may also be applied to part or all of the screw by passing water or other coolant through a passage at its centre, access being via a rotary union on the driven end of the screw.

The size of an extruder is defined by the nominal internal diameter of the barrel. Sizes range from about 25 mm for a laboratory machine, through 60~150 mm for most commercial product extrusions, up to 300 mm or more for homogenization during polymer manufacture. Modern thermoplastic extruders have screw length-to-diameter ratios of the order of 25 or more. Screws are often single start, that is, have one helical flight, with leads equal to their nominal external diameters. The channel depths and lengths of the three screw sections vary considerably, according to the type of polymer and application. An important characteristic of a screw is its compression ratio, one definition for which is the ratio between channel depths in the feed and metering sections. This ratio normally lies in the range 2~4, according to the type of material processed. Output rates obtainable from an extruder vary from about 10 kg/h for the smallest up to 5000 kg/h or more for the largest homogenisers. Screw-drive power requirements are usually of the order of 0.1~0.2 kW · h/kg.

Many modifications to the basic form of screw design are used, often with the aim of improving mixing. Another variant is the two-stage screw, which is effectively two screws in series. The decompression of the melt at the end of the first stage, where the screw channel suddenly deepens, makes it possible to extract through a vent any air or volatiles trapped in the polymer.

——Roger T. Fenner. Principles of Polymer Processing.
London: The Macmillan Press Ltd, 1979. 4-6

Words and Expressions

single-screw extrusion			单螺杆挤塑（成型）
finished product			成品
semifinished product			半成品
end product		n.	最终产品
homogenization	[hɔmeudʒeni'zeiʃən]	n.	均化（作用）
post-extrusion		n.	后挤出
granule	['grænju:l]	n.	颗粒，细粒
gravity feed			自动（重力）供料
hopper	['hɔpə]	n.	（进，加）料斗
crammer-feeding			填塞式供料
helical	['helikəl]	a.	螺旋状的
metering	['mi:təriŋ]	n.	计量，测量
perforate	['pə:fəreit]	v.	钻（打）孔
breaker plate			多孔板
screen pack			过滤网组
gauze	[gɔ:z]	n.	金属丝网
even	['i:vən]	vt.	使均匀（平衡）
thrust bearing			止推轴承，推力座
reduction gearbox			减速（齿轮）箱
radial	['reidiəl]	a.	径向的
clearance	['kliərəns]	n.	间隙
flight	[flait]	n.	螺齿
feed stock			原料
length-to-diameter ratio			长径比
compression ratio			压缩比
vent	[vent]	n.	排气孔，出口

UNIT 28 General Aspects of Polymer Degradation

What is "polymer degradation"? It is the collective name given to various processes which degrade polymers, i.e. deteriorate their properties or ruin their outward appearance.

Generally speaking, polymer degradation is a harmful process which is to be avoided or prevented. The operation which can be undertaken to inhibit or to retard degradation is called polymer stabilization. In order to do this suitably, with maximum efficiency, we must understand the mechanism of polymer degradation, we must identify the factors and stresses causing it, and we must know what factors affect polymer stability.

Sometimes, although not very often, polymer degradation may be useful. Depolymerization leading to high purity monomers may be exploited for practical production of such materials. Another important field in which degradation is desirable is the destruction of polymeric waste materials.

Degradation may happen during every phase of a polymer's life, i.e. during its synthesis, processing, and use. As mentioned before, even after the polymer has fulfilled its intended purpose, its degradation may still be an important problem. Depending on polymerization conditions, during polymer synthesis depolymerization may take place. From a practical point of view, it is important that the eventual fate of a polymeric material is often decided during its synthesis. ① Depending on the polymerization technology, varying amounts of "weak sites" may be built into the polymer which will later occasion its deterioration. For example, the presence of tertiary chlorines in PVC resulting from branch formation during polymerization may reduce PVC stability. During synthesis, various contaminants such as catalyzer residues or other polymerization additives may enter the polymer. The presence of impurities may be very decisive with respect to polymer life span. The amount of additives used during PVC polymerization techniques increase in the sequence block-suspension-emulsion; the stability of the polymer decreases in the same sequence.

During processing, the material is subjected to very high thermal and mechanical stress. These drastic stresses may initiate a variety of polymer degradation processes leading to a deterioration of properties even during processing. On the other hand, the damaging of the material may result in the introduction of various defects in the polymer which will work as degradation sources during its subsequent service life. ② For example, in the presence of oxygen traces during processing, carbonyl groups can be formed in polyolefins; these will later absorb UV light during outdoor applications and thus will function as built-in sensitizers for photodegradation processes. In PVC, a small amount of HCl may be eliminated from the polymer during processing. This alone does not influence the properties very much, but the resultant double bonds and the allyl-activated chlorines joined with them are very dangerous sources of PVC degradation.

The best known appearance of polymer degradation is connected with the use of these materials. Some kinds of polymer degradation, such as the outdoor aging of PVC roofs or the stiffening and discoloration of badly composed vinyl handbags, etc., are well known. The scale of polymer applications is, however, very broad; consequently, the stability requirements are highly diverse. In most cases there is a demand for a long service life; sometimes only a predicted (usually short) lifetime is required.

The problem of waste disposal is increasing with the use of increasing amounts of plastic materials. Organized recovery presently exists only in the case of production wastes inside polymer factories. In domestic garbage there is an every-increasing portion of plastic wastes, the destruction of which requires expensive equipment. During the burning of one pound of PVC, approximately 160 liters of HCl are evolved, which is very undesirable because of air pollution. Protection of the environment requires improved packaging materials which are "self-destructing", i. e. which will be degraded very rapidly when exposed to the effects of sunlight, humidity, and-finally-soil bacteria.

In summary, we can conclude that degradation plays an important role in every life phase of a polymer. Thus the importance of studying its mechanism, relationships, etc., is quite obvious.

——Tibor kelen. Polymer Degradation. New York: Van nostrand reinhold company, 1983. 1-3.

Words and Expressions

degradation	[,degrə'deiʃən]	n.	降解
stabilization	[,stəbilai'zeiʃən]	n.	稳定（化）
drastic	['dræstik]	a.	强烈的，激烈的
stress	[stres]	n.	应力
carbonyl	['kɑːbənil]	n.	羰基，碳酰
photodegradation	['fəutəudegrə'deiʃən]	n.	光降解

Notes

① "From practical point of view, ... decided during its synthesis." 译为："从实际的观点来看，聚合过程就可以决定其后聚合物的性能（这一事实）是不容忽视的。" "fate" 为命运。

② "...the damaging of the material may result in ... subsequent service life." 译为："…材料的破坏往往会在聚合物中引入各种缺陷，在聚合物材料以后的使用过程中，这些缺陷往往成为降解源。" 此处 "defect" 一词指材料 "结构上" 不完整的缺陷。

Exercises

1. *Translate the following paragraphs into Chinese*

There are various schemes to classify polymer degradation. Because of its complexity, with regard to both the causes and the response of the degradation, the classification is usually performed on the basis of the dominating features. One of the most frequent classifications has been based on the main factors responsible for degradations, such as thermal, thermo-oxidative, photo, photooxidative, mechanical, hydrolytic, chemical, and biological degradations; degradations by high energy radiation, pyrolysis and oxidative pyrolysis, etc.

Another possible classification is based on the main processes taking place as dominating event during degradation, such as random chain scission, depolymerization, cross-linking, side group elimination, substitution, reaction of side group among themselves, etc.

2. *List the examples of polymer degradations you've observed and classify them*

No.	polymer	phase of degradation	main phenomena	main causes
1	PVC	during processing	evolving of HCl	thermal
2				
3				
4				
5				
6				

3. *Put the following words into English*

降解　　　　　　交联　　　　　　解聚　　　　　　　　化学键断裂
生物降解　　　　光降解　　　　聚合物的稳定化　　　聚合物的稳定性

Reading Materials

Factors Affecting Polymer Stability

As previously mentioned, the chemical structure of the polymer is of primary importance in respect to its stability. The chemical composition (i.e. what kinds of chemical bonds in what sort of arrangement) is in itself a decisive factor. Bond energies between the same atoms are very different depending on the chemical groups to which the atoms belong. A few selected data are included in Table 28.1.

Table 28.1　Bond dissociation energies of various single bonds

bond broken A—B	bond dissociation energies/ (kJ/mol)	bond broken A—B	bond dissociation energies/ (kJ/mol)
C_2H_5—H	414	C_6H_5—H	431
n-C_3H_7—H	410	$C_6H_5CH_2$—H	347
t-C_4H_9—H	380	C_2H_5—CH_3	347
CH_2=$CHCH_2$—H	343	n-C_3H_7—CH_3	347
t-C_4H_9—CH_3	339	CH_2=$CHCH_2$—Cl	272
C_6H_5—CH_3	393	CH_3—F	451
$C_6H_5CH_2$—CH_3	301	C_2H_5—F	443
CH_3—Cl	351	HO—OH	213
C_2H_5—Cl	339	t-C_4H_9O—OH	150

Tertiary and allylic bonds are usually weaker than primary or secondary ones. In polymers consisting only of primary and secondary carbon atoms (e.g. PVC), the presence of such bonds is undesirable because these form weak sites which are very easy to attack. Processes leading to these bonds during polymerization (e.g. PVC branching or dehydrochlorination) are to be avoided.

The dissociation energies of the various bonds in the polymer may determine the course of degradation: the process always begins with the scission of the weakest available bond or with an attack at this site, and the first step usually determines the further direction of the process. Other components of the chemical structure, such as steric factors, stability of the intermediates, or the possibility of their resonance stabilization, may also have great influence on degradation. Such factors may even change the value of the bond dissociation energies.

Some comonomeric units incorporated in the copolymers may influence stability. Such units usually modify the application properties of the polymer, for example, the T_g or mechanical strength; they can, however, also improve the stability. Thus, the incorporation of a few percent of dioxolane units into polyformaldehyde greatly reduces its depolymerization because the dioxolane units inhibit the unzipping of formaldehyde units.

The presence of some comonomeric units, like the presence of certain additives or polymer blend components which are not chemically bonded to the polymer (although added in order to improve its properties), may decrease stability. For example, the rubber component of high impact polystyrene usually contains unsaturated bonds which are sources of various degradation reactions. High impact polystyrene products are improved materials from the viewpoint of their improved mechanical properties, but they are more sensitive to light or oxidation than polystyrene itself. An increasingly utilized method of improving stability is the chemical modification of polymers. Elimination of weak sites or substitution of labile groups by stable ones via grafting into the polymer may be very effective and useful especially when the substituents simultaneously improve other properties. Alkylation of the weak sites in PVC may result in internal plasticization of the material.

The tacticity of the polymer plays an important role in the degradation behavior. Atactic and isotactic polypropylene have very different oxidative stability (the isotactic one is much more stable). Syndiotactic PVC prepared at low temperatures has increased stability compared to the ordinary material produced at about 50℃. It is, however, very difficult to separate the effect of tacticity from that of the morphology of the material because a change in tacticity is usually connected with a change of morphology.

Physical and morphological factors may also influence polymer stability. It is well known that oxidation is always initiated in the amorphous phase of semicrystalline polymers and the propagation of the oxidation into the crystalline phase is a result of the destruction of the crystalline order. Thus, crystallinity is an important characteristic of the polymer from the viewpoint of stability.

The morphology of the material is decisive from the point of view of diffusion conditions. A compact material is usually more stable against oxidation because the diffusion of oxygen into the product is more difficult than with a material of loose structure. On the other hand, the facile diffusion of HCl evolved from PVC with a loose morphology reduces the autocatalytic character of the degradation which may lead to a catastrophic destruction in the case of compact and dense materials.

Similar to the internal chemical stresses already mentioned (weak sites, etc.), the internal mechanical stresses which are left in the material or introduced by finishing operation are very dangerous. Such stresses may serve not only as sources of later mechanical deterioration but also as initiators of, or assistants to, various chemical attacks. This is especially true in cases of stresses with long duration (the so-called stress corrosion of polymers).

The role of contaminants is quite obvious and has already been mentioned in connection with the synthesis of polymers. It is obvious that some additives intentionally present in the material, such as plasticizers or lubricants, influence the stability of the composite, especially if the oxidizability and biodegradability of such systems are higher than those of the polymer components. Once radicals are formed in the additive, they attack the polymer and vice versa; i. e. composites are sometimes less stable than their components. In special cases, additives are intentionally used to promote degradation of the composites (e. g. photosensitizers or plasticizers) which are specific culture media for bacteria in some rural and horticultural applications. On the other hand, additives may also stabilize the polymer: the use of antioxidants, photostabilizers, etc., which we will discuss later in detail, is based on this fact.

——Tibor kelen, Polymer Degradation. New York: Van Nostrand Reinbold Company, 1983. 4-8.

Words and Expressions

primary carbon			伯碳
steric (al)	[s'terik (əl)]	a.	立体的，立体化学的
labile	['leibail]	a.	易起物理（化学）变化的，不稳定的
plasticizer	[p'læstisaizə]	n.	增塑剂
horticulture	['hɔːtikʌltʃə]	n.	园艺（学）
photostabilizer	[fəutəu'steibilaizə]	n.	光稳定剂

UNIT 29 Synthetic Plastics

It would be difficult to visualise our modern world without plastics. Today they are an integral part of everyones lifestyle with applications varying from commonplace articles to sophisticated scientific and medical instruments. Nowadays designers and engineers readily turn to plastics because they offer combinations of properties not available in any other materials. Plastics offer advantages such as lightness, resilience, resistance to corrosion, colour fastness, transparency, ease of processing, etc., and although they also have their limitations, their exploitation is limited only by the ingenuity of the designer.

It is usual to think that plastics are a relatively recent development but in fact, as part of the larger family called "polymers", they are a basic ingredient of plant and animal life. Polymers are materials which consist of very long chain-like molecules. Natural materials such as silk, shellac, bitumen, rubber and cellulose have this type of structure. However, it was not until the 19th century that attempts were made to develop a synthetic polymeric material① and the first success was based on cellulose. This was a material called "Parkesine", after its inventor Alexander Parkes, and although it was not a commercial success it was a start and eventually led to the development of "Celluloid". This material was an important break-through because it became established as a good replacement for natural materials which were in short supply.

During the early twentieth century there was considerable interest in these new synthetic materials. Phenol-formaldehyde resin ("Bakelite") was introduced in 1905 and about the time of the second World War materials such as nylon, polyethylene and acrylic ("Perspex") appeared on the scene. Unfortunately many of the early applications for plastics earned them a reputation as being cheap substitutes. It has taken them a long time to overcome this image but nowadays the special properties of plastics are being appreciated which is establishing them as important materials in their own right.② The ever increasing use of plastics in all kinds of applications means that it is essential for designers and engineers to become familiar with the range of plastics available and the types of performance characteristics to be expected so that they can be used to the best advantage.③

The words "polymers" and "plastics" are often taken as synonymous but in fact there is a distinction. The polymer is the pure material which results from the process of polymerization and is usually taken as the family name for materials which have long chain-like molecules and this includes rubber. Pure polymers are seldom used on their own and it is when additives are present that the term plastic is applied. Polymers contain additives for a number of reasons. In some cases impurities are present as a result of the polymerization process and it may be uneconomic to remove these to get the pure polymer. In other cases additives such as stabilizers, lubricants, fillers, pigments, etc., are added to enhance the properties of the material.

There are two important classes of plastics:

(1) Thermoplastic materials. In a thermoplastic material the long chain-like molecules are held together by relatively weak Van der Waals forces. A useful image of the structure is a mass of randomly distributed long strands of sticky wool. When the material is heated the intermolecular forces are weakened so that it becomes soft and flexible and eventually, at high temperatures, it is a viscous melt. When the material is allowed to cool it solidifies again. This cycle of softening by heat and solidifying when cooled can be repeated more or less indefinitely and is a definite advantage in that it is the basis of most processing methods for these materials. It does have its drawbacks, however, because it means that the properties of thermoplastics are heat sensitive. A useful analogy which is often used to describe these materials is that like candle wax they can be repeatedly softened by heat and will solidify when cooled.

Examples of thermoplastics are polyethylene, polyvinyl chloride, polystyrene, nylon, cellulose acetate, acetal, polycarbonate, polymethyl methacrylate and polypropylene.

(2) Thermosetting materials. A thermosetting material is produced by a chemical reaction which has two stages. The first stage results in the formation of long chain-like molecules similar to those present in thermoplastics, but still capable of further reaction. The second stage of the reaction takes place during moulding, usually under the application of heat and pressure. The resultant moulding will be rigid when cooled but a close network structure has been set up within the material. During the second stage the long molecular chains have been interlinked by strong bonds so that the material cannot be softened again by the application of heat. If excess heat is applied to these materials they will char and degrade. This type of behaviour is analogous to boiling an egg. Once the egg has cooled and is hard, it cannot be softened again by the application of heat.

Since the cross-linking of the molecules is by strong chemical bonds thermosetting materials are characteristically quite rigid materials and their mechanical properties are not heat sensitive. Examples of thermosets are phenol formaldehyde, melamine formaldehyde, urea formaldehyde, epoxies and some polyesters.

——Crawford RJ. Plastics Engineering. Oxford: Pergamon Press, 1981. 75-79

Words and Expressions

visualise	['vizjuəlaiz]	vt.	想象，观察
integral	['intigrəl]	a.	必备的，构成整体所需要的
sophisticated	[sə'fistikeitid]	a.	复杂的，尖端的
resilience	[ri'ziliəns]	n.	弹性，回弹力
transparency	[træns'pɛərənsi]	n.	透明性，透光度
ingenuity	[indʒi'njuːəti]	n.	创造力，机敏
ingredient	[in'griːdiənt]	n.	组成部分，要素
shellac	[ʃə'læk]	n.	虫胶（漆）
bitumen	['bitjumin]	n.	沥青
parkesine	['pɑːkisain]	n.	硝化纤维素塑料
celluloid	['seljuloid]	n.	赛璐珞，假象牙
break-through		n.	突破，重要技术成就
phenol-formaldehyde resin			酚醛树脂
bakelite	['beikəlait]	n.	酚醛树脂，电木粉
acrylic	[ə'krilik]	a.	丙烯酸的
perspex	['pəːspeks]	n.	聚甲基丙烯酸甲酯，有机玻璃
synonymous	[si'nɔniməs]	a.	同义的
stabilizer	['steibilaizə]	n.	稳定剂
lubricant	['ljuːbrikənt]	n.	润滑剂
pigment	['pigmənt]	n.	颜料

Van der Waals force			范德瓦耳斯力
analogy	[ə'nælədʒi]	n.	比喻，类似
candle wax			烛用蜡
cellulose acetate			乙酸纤维素
polycarbonate	[ˌpɔli'kɑːbəneit]	n.	聚碳酸酯
acetal	['æsitæl]	n.	聚甲醛，缩醛

Phrases

(to) appear on the scene 出现，登场
on one's own 单独，独自
to the best advantage 以最好的方式
analogous to... 与…类似
in one's own right 凭自身的权利（资格）

Notes

① "it was not until the 19th century that attempts were made to develop a synthetic polymeric material…" 此句属一种强调句，这种强调句可以用于强调主语、宾语、表语、状语、介词短语、从句等，其基本结构为："It is（was）+ 被强调部分 + that…"。对于"not…until…"句式来说，在强调介词短语"until…"时，其句子的基本结构为："It is（was）not until that…"，故若将以上强调句还原成非强调结构的自然语序应为："Attempts were not made to develop a synthetic polymeric material until the 19th century…"。本句译文："一直到19世纪才进行了开发合成聚合材料的尝试。"

② "…nowadays the special properties of plastics are being appreciated which is establishing them as important materials in their own right." 句中"which"引出的定语从句是用来修饰主语"the special properties of plastics"的，因其较长，故将其放在句子后面，而在主语和这个定语从句之间插入了谓语，这样一来句子显得更紧凑一些。翻译时可采用主句和定语从句分译的方法。此句可译为："…现在，塑料的特性正在被人们认识到，而塑料凭其这些特性也正在确立它们作为重要材料的地位。"

③ "The ever increasing use of plastics in all kinds of applications means that it is essential for designers and engineers to become familar with the range of plastics available and the types of performance characteristics to be expected so that they can be used to the best advantage." 句中"means"是谓语，"that it is essential…"是宾语从句，"so that…"引出的是一个结果状语从句。本句可译为："塑料在各种应用中用途不断增加，这意味着对于设计师和工程师来说，熟悉可获得的塑料的范围和各种塑料的预期性能特点是十分必要的，以便以最好的方式利用它们。"

Exercises

1. *Complete the summary of the text. Choose no more than three words from the passage for each blank*

　　Nowadays plastics are ＿＿＿＿＿＿＿ of everyones lifestyle with a great variety of applications. Designers and engineers turn to plastics because they offer ＿＿＿＿＿＿＿ not available in any other materials.

　　Polymer materials are a ＿＿＿＿＿＿＿ of plant and animal life. However, it was not until the 19th century that attempts were made to develop a ＿＿＿＿＿＿＿ and the first success was based on cellulose. ＿＿＿＿＿＿＿ was introduced in 1905 and then many new plastics appeared on the scene. The ever increasing use of plastics means that it is essential for technologists to become familiar with plastics.

　　The polymer is the ＿＿＿＿＿＿＿ which results from the polymerization. When polymer contains ＿＿＿＿＿＿＿, the term plastics is applied. Additives are added to enhance the ＿＿＿＿＿＿＿ of the materials.

　　There are two classes of plastics: thermoplastics and thermosets. In a thermoplastic material the molecule structure is _____, therefore, thermoplastics can be softened by heat and will _____ when cooled repeatedly. A thermosetting material is produced by a chemical reaction which has two stages. A _____ structure has been set up within the material so that the material cannot be softened again by heat.

2. *Try to compare thermoplastics with thermosets and complete the following table*

plastics	thermoplastics	thermosets
structure		
intermolecular force		
when heated		
heat sensitive (yes/no)		
analogue		

3. *Put the following words into Chinese*

    visualise   ingredient   lubricant   polycarbonate
    integral   analogous   filler   epoxy
    sophisticated   stabilizer   pigment   polyester

4. *Put the following words into English*

    醋酸纤维素   透明性   酚醛树脂   弹性
    易加工性   聚氯乙烯   抗腐蚀性   创造力
    聚苯乙烯   固色性   赛璐珞   聚甲基丙烯酸甲酯

Reading Materials

Engineering Plastics

The Dictionary of Scientific and Technical Terms defines engineering plastics as those "plastics which lend themselves to use for engineering design, such as gears and structural members". This implies that engineering plastics can substitute for traditional materials of construction, particularly metals. The terms engineering plastics and engineering thermoplastics can be used interchangeably. In fact, it is convenient to restrict the term to plastics that serve engineering purposes and are processed by injection and extrusion methods. This excludes the so-called speciality plastics, e. g., fluorocarbon polymers and infusible film products such as polyimide film, and thermosets including phenolics, epoxies, urea-formaldehydes, and silicones. Many other polymers can be processed by injection molding and extrusion, but do not exhibit enough toughness, temperature resistance, or dimensional stability for engineering use. For the purposes of this article, the following definition is offered: engineering plastics are thermoplastics that maintain dimensional stability and most mechanical properties above 100℃ and below 0℃. This definition encompasses plastics which can be formed into functional parts that can bear loads and withstand abuse in temperature environments commonly experienced by the traditional engineering materials: wood, metals, glass, and ceramics. Generic resins falling within the scope of this definition include acetal resins, polyamides (nylons), polyimides, polyetherimides, polyesters, polycarbonates, polyethers, polysulfide polymers, polysulfones, blends or alloys of the foregoing resins, and some examples from other resin types. Thermoplastic elastomers and commodity resins, such as most styrenic resins, acrylics, polyolefins, polyurethanes, poly (vinyl chloride) s, and related chlorinated polyolefins, lose mechanical properties below 100℃. Exceptions include certain acrylonitrile-butadiene-styrene polymers (ABS) which maintain engineering properties in the vicinity of 100℃; some unreinforced nylons exhibit creep under load.

The development of engineering thermoplastic resins began with the pioneering work of Carothers on nylons at the Du Pont Co. in the 1930s and has continued with the recent introduction by General Electric Co. of its polyetherimide resin. Certain trends are discernible. First, the introduction of new polymers has slowed. Several new varieties of established polymer resins were introduced in the 1960s and 1970s, but polyetherimide resins are the only new generic polymers that have been commercialized in the 1980s. This is not surprising in light of the high risk associated with the introduction of new materials costing millions of dollars and years of development. However, many new blends of engineering materials, often tailored to specific applications, have been tried in the last few years. For these, only 2~4 yr are required for commercialization.

Plastics are either crystalline or amorphous. Acetals, most nylons, some polyesters, poly (aromatic sulfide), and polyetherketones are crystalline resins. In general, crystalline polymers do not transmit light, whereas amorphous polymers do. The mode of processing or treatment can alter the morphological form. However, a polymer is considered crystalline if that is the morphology which develops upon cooling the melt to ambient temperature. Amorphous polymers are defined similarly.

By far the most distinguishing properties between amorphous and crystalline polymers are organic solvent resistance and light transmission. In applications where solvent resistance is important, such as in a gear box exposed to oils and greases, a crystalline polymer is the likely choice. On the other hand, for filter bowls, glazing, etc., where clarity is required, an amorphous polymer might be preferred. When treatment of polymers results in morphological change, properties can change as well. For example, poly (ethylene terephthalate) (PET) normally solidifies from a melt as a partially crystalline, opaque solid. As a biaxially oriented film, the polymer is even more crystalline owing to heat-setting under restraint, yet it is also transparent.

Lubricity of crystalline polymers is usually higher than that of amorphous polymers. Excellent machinery parts are made from nylon-66 resins, e. g., gears, cams, wedges, and other components not requiring lubrication. Gears made of polyimide resin, on the other hand, do not exhibit this feature.

Crystalline polymers tend to be more stable dimensionally than amorphous materials. The closer packing and regular physical interactions of the crystalline species would be expected to yield less creep and susceptibility to deformation. Crystalline nylon expands and contracts with changes in ambient moisture. Thus, although crystalline polymers are dimensionally stable at a given rh, changes in humidity can result in warpage. This is important for large parts where dimensional change of a few micrometers per centimeter can result in serious deformation. In addition, crystalline polymers exhibit high mold shrinkage compared with amorphous materials. Mold shrinkages of most amorphous resins are ca 0.005 mm/mm, whereas crystalline resins shrink ca 0.015 mm/mm.

Polymers with differing morphologies respond differently to fillers and reinforcement. In crystalline resins, heat-distortion temperature (HDT) increases as the aspect ratio and amount of filler and reinforcement are increased. In fact, glass reinforcement can result in the HDT approaching the melting point. Addition of fillers, however, interrupts the physical interactions, and certain properties, such as impact strength, are reduced. Amorphous polymers are much less affected.

Engineering thermoplastics are priced between the very expensive resins and the high volume, low priced commodities, e.g., poly (phenylene oxide)-polystyrene alloys at $2/kg to polyether-etherketones at $60/kg. Many speciality and commodity resins are addition polymers. Their prices are primarily governed by raw material costs. Thus, the raw material for fluoropolymers is much more expensive than the olefins used for polyethylene (PE), polystyrenes (PS), acrylics, vinyl choride polymers, etc. Engineering polymers tend to be composed of aromatic monomers, except for acetals and nylons, linked by condensation (ester, carbonate, amide, imide), substitution (sulfide), or oxidative coupling (ether). Monomer costs and complexity of polymerizations account for higher prices. Larger volumes of production take advantage of economics of scale.

——Donald C Clagett. Encyclopedia of Polymer Science and Engineering. Vol 6. Editor-in-Chief Jacqueline I Kroschwitz. 2nd ed. New York: John Wiley & Sons, 1986. 94-97

Words and Expressions

gear	[giə]	n.	齿轮
polyimide	[pɔli'imid]	n.	聚酰亚胺
phenolics	[fi'nɔliks]	n.	酚醛塑料（树脂）
urea-formaldehyde		n.	脲甲醛树脂
silicone	['silikən]	n.	（聚）硅氧烷，有机硅树脂
encompass	[in'kʌmpəs]	vt.	包含（括，围）
polyetherimide	[pɔli'i:θə'imid]	n.	聚醚酰亚胺
polysulfide	[pɔli'sʌlfaid]	n.	聚硫化物，多硫化合物
polysulfone	[pɔli'sʌlfɔn]	n.	聚砜
foregoing	[fɔ:'gəuiŋ]	a.	前面的，上述的
thermoplastic elastomer			热塑性弹性体
acrylonitrile-butadiene-styrene polymer (ABS)			丙烯腈-丁二烯-苯乙烯聚合物
vicinity	[vi'siniti]	n.	附近
yr(year)		n.	年
poly(aromatic sulfide)		n.	聚芳烃硫化物
ambient	['æmbiənt]	a.	周围的，环境的
grease	[gri:s]	n.	油脂
filter bowl			滤罩
glazing	[g'leiziŋ]	n.	装玻璃，抛光
clarity	['klæriti]	n.	透明（光）度
poly(ethylene terephthalate) (PET)		n.	聚对苯二甲酸乙二醇酯
opaque	[əu'peik]	a.	不透光［明］的，无光泽的
biaxially	[bai'æksiəli]	adv.	二（双）轴向地
heat-setting		n.	热定形，热固化
cam	[kæm]	n.	凸轮
wedge	[wedʒ]	n.	楔（块，形物）
rh(relative humidity)			相对湿度
warpage	['wɔ:peidʒ]	n.	翘曲，扭曲
mold shrinkage			（脱）模后收缩
ca = circa	['sə:kə]	prep.	大约
heat-distortion temperature (HDT)			热变形温度
poly(phenylene oxide)-polystyrene alloy		n.	聚苯氧-聚苯乙烯合金
polyetheretherketone	[pɔli'i:θəi:θə'ki:tən]	n.	聚醚醚酮
fluoropolymer	[fluərə'pɔlimə]	n.	含氟聚合物
oxidative coupling			氧化偶合（作用）

UNIT 30 Synthetic Rubber

The earliest recorded accounts of the production of synthetic rubber by polymerization refer to isoprene (2-methyl-1,3-butadiene), a substance which is closely related to the repeat unit in the molecule of natural rubber. Williams appears to have been the first to isolate this compound from the products obtained by the destructive distillation of natural rubber in the absence of air, and to show that it becomes viscous in the presence of air.① Furthermore, when this viscous liquid was heated, it was found to change into a spongy, rubbery substance. That isoprene could be converted into a rubbery solid by treatment with hydrochloric acid was demonstrated by Bouchardat and by Tilden. The first patent to be granted for the manufacture of synthetic rubber was to Matthews and Strange in 1910 for the production of a rubbery substance from isoprene by treating the latter with sodium metal.

Considerable interest was shown in the production of synthetic rubber from dimethylbutadiene in Germany during World War Ⅰ. Cut off from supplies of natural rubber by the British blockade, it was necessary to seek alternative materials. The product obtained by polymerization of dimethylbutadiene was known as "methyl rubber". During the late 1920s, interest shifted from dimethylbutadiene to 1,3-butadiene as a monomer because a more satisfactory rubber was obtained from the latter and because it was more readily (available) than isoprene. Polymerization was effected using sodium metal as catalyst. From this fact originated the word "Buna" as a generic name for synthetic rubbers produced in Germany at that time and subsequently.②

Two important developments occurred in the U.S.A. in the years immediately following World War Ⅰ. The first was the use of butadiene as a monomer. The second was the introduction of the technique known as emulsion polymerization for converting the monomer to synthetic rubber. Until comparatively recently, emulsion polymerization has been the principal process for producing synthetic rubbers from their monomers. The most significant early event in the development of special-purpose synthetic rubbers may well have been the announcement in 1931 by the Du Pont Company that they had succeeded in developing a new synthetic rubber which they called "Neoprene". This synthetic rubber is produced by the polymerization of 2-chloro-1,3-butadiene, commonly known as "chloroprene". It is the presence of this chlorine atom in each repeat unit of the rubber molecule which gives the product its unusual properties,③ such as moderate resistance to swelling in hydrocarbon oils and resistance to deterioration by heat and ozone. Polysulphide rubbers were also developed around the early 1930s.

Copolymers of butadiene with styrene and with acrylonitrile were investigated in Germany in the early 1930s. Emulsion-polymerized copolymers of styrene and butadiene were found to have better mechanical properties than butadiene homopolymers. Copolymers of acrylonitrile and butadiene were found to have interesting oil-resistance properties. The rubbers obtained from butadiene and styrene were designated as "Buna S" and those from butadiene and acrylonitrile as "Buna N".

The outstanding development in synthetic rubber manufacture during World War II was undoubtedly that of emulsion-polymerized styrene-butadiene rubber. This development took place in both Germeny and the U.S.A.. One other important synthetic rubber development must also be mentioned. This concerns the so-called "butyl" rubber, a copolymer of isobutene with a minor amount of isoprene. This rubber was made by a type of polymerization (cationic polymerization). Butyl rubber was first announced in 1940. It has relatively low permeability to gases and displays good resistance to cracking by ozone and to embrittlement on ageing.

A series of important developments in the manufacture of emulsion polymerized styrene-butadiene rubbers has been exploited since the end of World War II. Of these, the first and most important was the discovery that the emulsion polymerization reaction could with advantage be carried out at much lower temperatures (5°C) than hitherto (50°C). Special types of polymerization initiator are required for this reaction. Perhaps the most exciting development in the area of general-purpose synthetic rubbers is that of the stereospecific polyisoprenes—the "synthetic natural" rubbers. In 1956, two methods were disclosed for the polymerization of isoprene to essentially *cis*-1,4-polyisoprene. In one of them, polymerization is effected in solution by means of lithium metal or by alkyllithiums. The reaction is an anionic polymerization. The second method uses a catalyst of the so-called "Ziegler-Natta" type, that is, a complex formed by reacting an organometallic compound with a transition-metal compound. In addition, the development of ethylene-propylene copolymers which can be used as general-purpose synthetic rubbers is yet another important landmark in the story of the development of synthetic rubbers.

There are three other developments which must be mentioned in this general review. The first concerns that of materials which are often known as "thermoplastic rubbers". One commercially-important type is a styrene-butadiene-styrene "block" copolymer. The second development is that of powdered rubbers. These are rubbers in a finely-powdered form which facilitates mixing of the rubber with other compounding ingredients. The third development is that of fluid rubbers. These are high-viscosity liquids which are sufficiently fluid to be capable of being cast into moulds,④ and which can then be vulcanised to give solid elastomeric materials.

——Blackey D C. Synthetic Rubbers. Barking: Applied Science Publishers Ltd., 1983. 17-30

Words and Expressions

isoprene	['aisəupri:n]	n.	异戊二烯
2-methyl-1,3-butadiene		n.	2-甲基-1,3-丁二烯
destructive distillation			干馏
spongy	[spʌndʒi]	n.	海绵状的
dimethylbutadiene		n.	二甲基丁二烯
blockade	[blɔ'keid]	n.	封锁
methyl rubber			甲基橡胶
originate	[ə'ridʒineit]	v.	起源,首创
Buna	['bju:nə]	n.	丁钠橡胶
special-purpose		n.	特殊用途的,专用的
the Du Pont Company			(美国的)杜邦公司
neoprene	['ni:əpri:n]	n.	氯丁(二烯)橡胶
chloroprene	['klɔrəpri:n]	n.	氯丁二烯

swelling	['sweliŋ]	n.	溶胀
hydrocarbon oil			烃油
deterioration	[ditiəriə'reiʃən]	n.	变质，变坏
polysulphide rubber			聚硫橡胶
oil-resistance		n.	耐油性
Buna-S		n.	丁苯橡胶
Buna-N		n.	丁腈橡胶
styrene-butadiene rubber			丁苯橡胶
butyl rubber			丁基橡胶
isobutene	[aisəu'bjuti:n]	n.	异丁烯
cracking	['krækiŋ]	n.	龟裂，裂纹
embrittlement	[em'britlmənt]	n.	脆裂，脆性
hitherto	['hiðətu:]	adv.	迄今，向来
stereospecific	[stiəriəuspi'sifik]	a.	立体定向的，有规立构的
general-purpose		a.	通用（型）的
cis-1,4-polyisoprene		n.	顺（式）聚异戊二烯
alkyllithium	[ælki'liθiəm]		烷基锂
organometallic compound			有机金属化合物
transition-metal compound			过渡金属化合物
facilitate	[fə'siliteit]	vt.	使容易，便于
vulcanise	['vʌlkənaiz]	v.	硫化，硬化
elastomeric	[i:'læstəumərik]	a.	弹性（体）的

Phrases

an account of 关于…的说明
cut off 截断，切断
originate from 起源于

appear to be 可认为是，似乎是
shift from …to … 从…转（移）到…
succeed in (+ing) 成功地（…），得以（…）

Notes

① "Williams appears to have been the first to isolate this compound from the products obtained by the (destructive) distillation of natural rubber in the absence of air, and to show that it becomes viscous in the presence of air." 句中 "appears to have been the first" 是系动词加表语构成的复合谓语。后面的两个不定式短语 "to isolate…, and to show…" 是 "the first" 的定语，这部分比较长，也比较复杂。翻译时可采用增加句量分译的方法。此句可译为："威廉斯可以说是在这方面取得成果的第一人。他在隔绝空气的条件下从干馏天然橡胶所得的产物中分离出了这种化合物，并且证实这种化学物质在空气中会变成黏稠状的。"

② "From this fact originated the word 'Buna' as a generic name for synthetic rubbers produced in Germany at that time and subsequently." 此句是一种主谓倒置的全部倒装结构。句中主语 "the word 'Buna' as …" 较长，谓语是 "originated"，没有宾语。为了强调状语，故将状语 "From this fact" 放在句首，接着把谓语放在主语前面，这样使句子比较均衡，也把状语摆在较突出的地位，并使整个句子与前面的句子联系得更紧密。此句可译为："丁钠橡胶这个词作为德国在那时及其以后生产的合成橡胶的通用名称就是源于这一事实。"

③ "It is the presence of this chlorine atom in each repeat unit of the rubber molecule which gives the product its unusual properties." 这是一个强调句型（把句中的 it is 和 which 去掉，剩下的恰好是一个完整的句子）。这里强调的是主语。此句可译为："正是氯原子在橡胶分子每个重复单元中的存在赋予了它的产品独特的性能。"

④ "being cast into moulds" 是动名词短语，其中 "being cast" 是动名词被动式形式（cast 的过去式和过去分词均与原形相同）。

Exercises

1. *Complete the summary of the text. Choose no more than three words from the passage for each blank*

   In 1860 Williams obtained _____ by destructive distillation of natural rubber. By 1887 Bouchardat, Tilden and Wallach had all converted isoprene back into a _____. Matthews and Strange were granted the first patent for the manufacture of _____ in 1910.

   During World War I, German scientists developed _____ rubber. In the late 1920s, Buna rubbers were produced.

   After World War I in the U.S.A. butadiene was used as a monomer and polymerization was introduced. In 1931 the Du Pont Company had succeeded in developing _____. Polysulphide rubbers were also developed around the early 1930s.

   Copolymers of butadiene with styrene and with acrylonitrile were investigated in Germany in the early 1930s. The rubbers obtained from them were respectively designated and _____.

   During World War II styrene-butadiene rubbers were developed outstandingly in Germany and the U.S.A.. One other important synthetic rubber development concerns _____.

   A series of important developments in the manufacture of synthetic rubber has been exploited since the end of World War II, such as low temperature emulsion polymerization, stereospecific _____ and ethylene-propylene copolymers.

   Three other developments are _____ rubbers, _____ rubbers and _____ rubbers.

2. *Make a sentence out of each group of words*

   (1) rubbery...could...into...converted...solid...be...a...isoprene

   (2) product...the...dimethyl-butadiene...by...obtained...of...polymerization...as...known...methyl rubber...was

   (3) synthetic...process...the...rubbers...has...for...polymerization...been...emulsion...producing...principal

   (4) developments...other...be...three...must...which...are...there...mentioned

3. *Put the following words into Chinese*

isoprene	2-methyl-1,3-butadiene	methyl rubber	butyl rubber
chloroprene	polysulphide rubber	Buna-S	*cis*-1,4-polyisoprene
neoprene	styrene-butadiene rubber	Buna-N	stereospecific

4. *Put the following words into English*

干馏	变质	龟裂	通用的
海绵	耐油性	脆性	专用的
溶胀	透气性	硫化	弹性体的

Reading Materials I

Nature of Elastomers

ASTM D1566—90 defines elastomers as "macromolecular material that returns rapidly to approximately the initial dimensions and shape after substantial deformation by a weak stress and release of the stress." It also defines a rubber as: "material that is capable of recovering from large deformations quickly and forcibly, and can be, or already is, modified to a state in which it is essentially insoluble (but can swell) in solvent, such as benzene, methyl ethyl ketone, and ethanol toluene azeotrope."

A rubber in its modified state, free of diluents, retracts within 1 min to less than 1.5 times its original length after being stretched at room temperature (20 to 27℃) to twice its length and held for 1 min before release. More specifically, an elastomer is a rubberlike material that can be or already is modified to a state exhibiting little plastic flow and quick and nearly complete recovery from an extending force. Such material before modification is called, in most instances, a raw or crude rubber or a basic high polymer and by appropriate processes may be converted into a finished product.

When the basic high polymer is converted (without the addition of plasticizers or other diluents) by appropriate means to an essentially nonplastic state, it must meet the following requirements when tested at room temperature (15 to 32°C):

(1) It is capable of being stretched 100 percent;
(2) After being stretched 100 percent, held for 5 min, and then released, it is capable of retracting to within 10 percent of its original length within 5 min after release.

The rubber definition with its swelling test certainly limits it to only the natural latex-tree source, whereas the elastomer definition is more in line with modern new synthetics.

——Joseph F Meier. Handbook of Plastics, Elastomers and Composites. Edited by Charles A Harper. 2nd ed. New York: McGraw-Hill Inc, 1992. 1.84-1.87

Words and Expressions

ASTM	美国材料试验学会
methyl ethyl ketone	丁酮（甲基乙基酮）
ethanol toluene azeotrope	乙醇甲苯恒沸物（法）
crude rubber	生胶，天然橡胶

Reading Materials II

Thermoplastic Elastomers (TPEs)

Though not a single family of materials, thermoplastic elastomers have a set of properties in common. They can be processed by conventional thermoplastic extrusion and molding techniques but function much like thermoset rubbers.

The earliest commercial TPEs, introduced in the 1950s, were the thermoplastic polyurethanes (TPUs). Since that time, other types of materials have been commercialized, among them styrenic block copolymers, copolyesters, olefin blends, and rubber olefin alloys.

TPUs are on the high end of the price and property range of thermoplastic elastomers. They can be processed easily at low melt temperatures (under 232°C) but must be thoroughly predried. Because of their natural adhesive properties, mold releases usually are required.

TPUs are extruded into hoses and tubing for automotive, medical, and electronics applications. Blown film is used for meat wrapping. Injection-molded TPU parts are used in auto exteriors, shoes and boots, drive wheels, and gears. TPUs can be blended with PVC, styrenics, nylons, or polycarbonate to achieve a wide range of properties.

Styrenic block copolymers are the most commonly used TPEs. Their structure normally consists of a block of rigid styrene on each end with a rubbery phase in the center, such as styrene/butadiene/styrene (SBS), styrene/isoprene/styrene (SIS), styrene/ethylene-butylene/styrene (SEBS), and styrene/ethylene-propylene/styrene (SEPS).

Styrenic TPEs can be processed via conventional extrusion and molding techniques. With a hardness range from 28 to 95 Shore A, they can be made into a wide variety of products. Wire and cable jacketing, medical tubing and catheters, shoe soles, and flexible automotive parts are some examples.

Compatibility with many other polymers has led to the use of styrenic TPEs as impact modifiers and as compatibilizers for resin blends. They also are used as tie layers in coextruded products.

Olefinic TPEs, or TPOs as they are typically called, are blends of polypropylene with rubber (EPDM or EP) and polyethylene. They are characterized by high impact strength, low density, and good chemical resistance. TPOs can be easily processed by extrusion, injection, and blow molding.

Gas-phase polyolefin producers can make in-reactor TPOs with properties similar to those made by compounding. The in-reactor products offer cost advantages for large-volume applications. The largest single market for TPOs is in automotive exterior parts. Paintable formulations are used to mold bumper, fascia, air dams, and rub strips. Weatherable grades with molded-in accent colors are top-coated with a clear finish.

Olefin blends with vulcanized rubber (partially or fully crosslinked) form a special category of TPE. These materials can cover the broad property range from thermoplastic to thermoset rubber. They process easily in conventional extrusion and molding equipment and are replacing thermoset rubbers in applications such as hose and tubing, cable covering, gaskets and seals, and sheeting. Special grades are available for medical and food contact applications.

Recently, olefinic TPEs have been developed that use proprietary olefinic materials in place of vulcanized rubber as the soft phase. They are made by special blending and crosslinking methods.

Polyester TPEs are high-strength elastomeric materials. They are block copolymers with hard and soft segments providing strength and elasticity. The primary characteristic of polyester TPEs is their ability to withstand repeated flexing cycles. They retain their high impact strength at low temperatures and may have heat resistance up to 149℃. Polyester TPEs process well (they should be predried) by extrusion and molding. Applications include automotive parts, sporting goods, and geomembrane sheeting.

—— Michael L Berins. SPI Plastics Engineering Handbook. 5th ed. New York: Van Nostrand Reinhold, 1991. 72-73

Words and Expressions

copolyester	[kəupɔli'estə]	n.	共聚多酯
predry	[pri:'drai]	v.	预干燥
mold release			脱模剂
hose	[həuz]	n.	胶管
tubing	['tju:biŋ]	n.	胶管,内胎
wrapping	['ræpiŋ]	n.	包卷
auto	['ɔ:təu]	n.	汽车
shoe	[ʃu:]	n.	刹车块,轮(外)胎
boot	[bu:t]	n.	(汽车后部)行李箱

drive wheel			主动轮
Shore A		n.	邵氏硬度 A
jacketing	['dʒækitiŋ]	n.	保护层
catheter	[kə'θi:tə]	n.	导管
TPO		n.	热塑性聚烯烃
EPDM		n.	三元乙丙橡胶
EP		n.	二元乙丙橡胶，环氧树脂
bumper	['bʌmpə]	n.	保险杠
fascia	['feiʃə]	n.	（汽车）仪表盘
air dam			（汽车）导风板
rub strip			防擦挡条
weatherable	[we'ðərəbl]	a.	耐气候（老化）的
top coat			面漆，表面涂层
finish	['finiʃ]	n.	保护层，漆面
gasket	['gæskit]	n.	密封圈，垫圈（片）
proprietary	[præ'praiətəri]	a.	有专利权的
elasticity	['elæs'tisiti]	n.	弹性，弹力
sporting goods			体育用品

UNIT 31　Structure of Fiber-forming Polymers

Textile fibers are solid organic polymers distinguishable from other polymers by their physical properties and characteristic geometric dimensions. The physical properties of textile fibers, and indeed of all materials, reflect molecular structure and intermolecular organization. The ability of certain polymers to form fibers is attributed to several structural features at different levels of organization rather than to any one particular molecular property.

Fiber structure is described at three levels of molecular organization, each relating to certain aspects of fiber behavior and properties. Of first importance is the organochemical structure defining the chemical composition and molecular structure of the repeating unit in the polymer, and also the nature of the polymeric link. This primary level of molecular structure is directly related to chemical properties, dyeability, moisture sorption, swelling characteristics, and indirectly related to all physical properties. The macromolecular level of structure describes the chain length, length distribution, and stiffness, as well as molecular size and shape. The supramolecular organization is the arrangement of the polymer chains in three-dimensional space. The physical properties of fibers are strongly influenced by the organization of polymeric chains into crystalline and noncrystalline domains, and the disposition of these domains with respect to each other. This morphology is quite complex, with fibrils, microfibrils, and similar structural subunits frequently surrounded by a matrix material in a composite configuration.

All polymeric fibers that are useful in textile applications are semicrystalline, irreversibly oriented polymers, i.e. the polymeric chains are partially ordered into regions or domains with near-perfect registry where the laws of X-ray diffraction are obeyed. In other regions of the fiber, the molecular chains or segments of chains are not perfectly ordered and may approach random coil configurations. This is a simple statement of the two-phase crystalline-amorphous model of the structure of semicrystalline polymers that has been used to describe fiber structure and to interpret fiber properties.

Besides, a lot of other models picture a fiber as a polymeric substance with a high degree of three-dimensional structural regularity, and lead to the concept of the degree of crystallinity, i.e., the fractional crystalline content of a partially crystalline polymeric material. ①

The requirement of at least partial crystallinity limits the number of polymers suitable for fiber formation. To a large extent, a "fiber forming" polymer, is actually a "crystallizable" polymer. Some structural characteristics of polymers such as regularity, chain directionality, single-chain conformation and chain stiffness allow them to crystallize under appropriate conditions.

Attention must also be focused on the noncrystalline domains. Many important properties of fibers can be directly related to the noncrystalline or amorphous regions. For example, absorption of dyes, moisture, and other penetrants occurs in these regions. These penetrants are not expected to diffuse into the crystalline domains, although adsorption on crystallite surfaces has been postulated. The extensibility and resilience of fibers is also directly associated with the noncrystalline regions.

Fiber structure is a dual or a balanced structure. Neither a completely amorphous structure, nor a perfectly crystalline structure, would provide the balance of physical properties required in fibers. The formation and processing of fibers is designed to provide an optimal balance in terms of both structure and properties.

——Rebenfeld L. Encyclopedia of Polymer Science and Engineering. Vol 6. Editor-in-Chilf J I Kroschwitz. 2nd ed. New York: John Wiley & Sons, 1986. 662-664

Words and Expressions

dyeability	[,daiə'biliti]	n.	可染性,染色度
dye	[dai]	n.	染料
		v.	染色
disposition	[,dispə'ziʃən]	n.	陈列;布置
oriented	['ɔːrientid]	a.	定向的,取向的
penetrant	['penitrənt]	n.	渗透剂,渗入料
extensibility	[iks,tensə'biliti]	n.	伸长率,可延展性
postulate	['pɔstjuleit]	v.	假设,以…为前提
dual	['djuː(ː)əl]	a.	双重的
moisture absorption			吸湿性

Phrases

be attributed to 由于,来源于	rather than 而不是
(be) associated with 与…有关系	

Notes

① "…and lead to the concept of the degree of crystallinity, i. e., the fractional crystalline content of a partially crystalline polymeric material." 译为:"…而且引出了结晶度的概念,即在部分结晶聚合物中结晶区所占比例。"

Exercises

1. *Match each word on the left with its correct Chinese translation*

    noncrystalline   硬度
    textile   半结晶的
    dyeability   卷绕
    stiffness   小纤维
    fibril   回弹性
    semicrystalline   非结晶的
    coil   可染色性
    resilinece   纺织品

2. *Complete the following according to the text*

    (1) Usually, fiber structure in molecular organization is considered at three levels: _____, _____ and _____ level.

    (2) All polymeric fibers that are useful in textile applications are _____, irreversibly _____ polymers.

(3) Fiber structure is not a completely _____ structure, nor a perfect one, but a balance of them.

Reading Materials

Biomedical Applications of Fibers

Fibers have been used for many years in medical applications such as dressings, sanitary pads, tapes, and operating-room apparel. Applications of engineering fibers include sutures, surgical implants, prosthetic devices, and medical equipment. Engineering fibers are also used for separations and purification such as kidney dialysis, and they will play important roles in emerging biotechnology applications such as substrates for large-scale cell culture. These fiber aid in maintaining and improving the quality of life.

1. Fiber properties

Fibers possess certain advantages applicable to biomedical uses. The surface-to-volume ratio is very large, which is important in surface-dominated applications such as separations. Engineering fiber can also be processed into knitted, woven, or hollow structures that provide the varying degrees of porosity and strength needed in biomedical applications such as synthetic blood vessels, artificial ligaments, and controlled drug release.

Many engineering fibers are used within the body because of their good mechanical properties, long-term stability, and nontoxic character. These fibers also are biocompatible; that is, they do not produce any significant changes in the surrounding tissue over long periods of time. Biocompatibility of the fibers depends on the type of polymer from which they are made, on processing additives, and on the surface properties of the fiber. Woven or knitted fibers may approximate the stress-strain behavior of natural tissue more closely than other structures and thus offer greater compatibility. This is of particular importance in applications such as the replacement of natural blood vessels.

2. Fiber types

Both natural and synthetic engineering fiber are used in the biomedical field (Table 31.1). Natural fibers such as silk and catgut (collagen) have been used by physicians for many years. However, the introduction of synthetic polymers has brought about the increased use of fibers in biomedical applications. Polyester and nylon fibers are used in a large number of medical applications. Recently, higher performance fibers such as carbon have been evaluated, particularly in applications requiring high strength and improved tissue compatibility.

Table 31.1 Engineering fibers used in biomedical applications

type	examples
natural fibers	collagen, silk, cellulose, chitin
synthetic organic fibers	polyester, polyamide, polypropylene, polyurethane
	polytetrafluoroethylene, polyethylene, poly (vinyl alcohol)
	polyacrylonitrile, poly (glycolic acid), poly (lactic acid)
	regenerated cellulose, polydimethylsiloxane, aramid
inorganic fibers	carbon, graphite, alumina, glass

3. Implants

Engineering fibers are also used in surgical implants. One application is in synthetic replacements for diseased or nonfunctioning blood vessels, such as segments of the aorta or other large-diameter arteries. Polyester fiber and expanded polytetrafluoroethlene (PTFE) are currently the synthetic materials of choice for these applications. Polyester fibers produced from poly (ethylene terephthalate) is knitted or woven into a porous tube with a smooth, lightly napped surface. The PTFE is produced as a continuous microporous tube with a smooth, noncrimped flow surface.

Tensile strength, long-term durability, biocompatibility, and a controlled degree of porosity are requirements of the woven or knitted fiber tubing used in arterial replacements. The structure must have adequate strength to resist longitudinal and transverse tension, to withstand the intraluminal blood pressure, and to be suturable. These properties must be maintained sufficiently to prevent rupture or aneurysm formation for many years after implantation. The fibrous structure must also have sufficient porosity to allow tissue in-growth and the formation of a thin fibrin-based thromboresistant neointima on the inner surface of the tubing. On the other hand, the pores must be small enough to prevent the outflow of blood. Finally, the fiber must not elicit an adverse tissue response that would lead to rejection.

Whereas structures based on polyester fibers and PTFE have been acceptable in large-diameter, high-blood-flow applications, synthetic substitutes for smaller-diameter arteries require new elastomeric fibrous structures that are more thromboresistant.

Damaged ligaments and tendons may be replaced by orthopedic implants produced from engineering fibers. Natural ligaments and tendons are made up primarily of collagen fibers; ligaments connect two or more bones whereas tendons connect muscle to bone. Engineering fibers that are candidate replacements for these structures must approximate their mechanical and biological properties. Carbon, polyester, polypropylene, and polytetrafluoroethylene fibers are currently being investigated.

Carbon fibers have attracted interest for both ligament and tendon replacements. They can serve as scaffolds upon which new, oriented collagenous fibrous tissue, which ultimately functions as the replacement, grows. The fibers have been coated with poly (lactic acid) to contain migration of carbon fragments produced by premature mechanical degradation from the implant site. In addition to the excellent tissue-compatibility properties, other important properties of carbon fiber are its high ultimate tensile strength (2.6 GPa) and high modulus of elasticity (260 GPa). Carbon fibers, however, are brittle and therefore require particular care in surgical installation.

Polyester fibers have also been used in tendon implants. Composites of carbon fiber with polysulfone resin and carbon fiber-reinforced carbon are being evaluated in orthopedic devices such as artificial hips. The potential advantage is a closer match to the modulus of bone.

——Dhingra A K, Lauterbach H G. Encyclopedia of Polymer Science and Engineering. Vol 6. Editor-in chilf kroschwitz J I. 2nd ed. New York: John Wiley & Sons, 1986.762-763

Words and Expressions

suture	['sju:tʃə]	n.	缝线
		v.	缝合
surgical implants			外科手术植入物
kidney dialysis			肾透析
blood vessels			血管
catgut	['kætgʌt]	n.	肠线
tissue-compatibility		n.	组织相容性
biocompatibility	['baiəukəm,pætə'biliti]	n.	生物相容性
retention	[ri'tenʃən]	n.	停滞，固定
aorta	[ei'ɔ:tə]	n.	主动脉
longitudinal	[,lɔndʒi'tju:dinl]	a.	经度的，纵向的
transverse	['trænsvə(:)s]	a.	纬度的，横向的
intraluminal	[,intrə'lju:minl]	a.	管腔内的
aneurysm	['ænjuərizəm]	n.	动脉瘤
in-growth			内部生长
tendon	['tendən]	n.	腱
orthopedic	[,ɔ:θəu'pidik]	a.	整形的

UNIT 32 Matching Adhesive to Adherend

When two materials are bonded, the resultant composite has at least five elements: adherend No. 1/interface/adhesive/interface/adherend No. 2. The adhesive and adherend must be compatible, if their union is to last.

The strength of the adhesive joint will be the strength of its weakest member. If one of the adherends is paper, excessive stress will usually result in a "paper tear". With stronger substrates, however, the failure will be either adhesive at an interface or cohesive within the glue. Failure will not be at an interface if the adherend surface has been properly prepared and the adhesive wets the adherend and is otherwise appropriate. In other words, the adhesion between glue and substrate should be greater than the cohesion within the glue line. This will occur provided the combining of adhesive and adherend has caused a decrease in free energy, and provided also that excessive strains are not built up when the adhesive sets. ①

Lets us consider the latter requirement first. Adhesives usually shrink as they harden (Inorganic cements are exceptions). Polymerization, the loss of solvent, even the cooling of a hot-melt may cause the glue line to contract. Strains are set up which induce the adhesive to pull away from the substrate. In addition, strains are produced when the adhesive joint is flexed. Various remedies may lessen the danger of failure from these causes:

(1) Choose low-shrinking resins, e. g. epoxies rather than unsaturated polyesters.

(2) Choose adhesives that are less rigid than the adherends; otherwise flexing will cause a concentration of stress in the glue line (However, excessive flexibility in the adhesive may be accompanied by low cohesive strength).

(3) Keep the glue line as thin as possible, consistent with the smoothness of the adherends, if the stresses are chiefly tensile. But porous adherends require the application of sufficient adhesive to avoid a "starved glue line". If the joint is to be exposed to considerable shear stress, the glue line should be somewhat thicker.

(4) Incorporate inert and preferably inorganic fillers.

(5) After applying the adhesive to an impervious substrate, evaporate water or solvents thoroughly before mating with a second impervious adherend.

Turning now to the free energy requirements, let us examine the types of bonds that may exist between adhesive and adherend. These chemical bonds may be either primary or secondary.

Primary bonds include electrovalent, covalent, and metallic bonds. *Electrovalent* or *heteropolar* bonds may be a factor in protein adhesives. *Covalent* or *homopolar* bonds may play a part in some finishing treatments for fiber glass. The *metallic* bond is formed by welding, soldering, and brazing. The inorganic materials for these purposes are essentially high temperature thermoplastic adhesives, but are outside the scope of this paper.

By far the most important of the adhesive bonds are the *secondary* or *Van der Waals'* bonds that give rise to attraction between molecules. Most significant of these are the *London* or *dispersion forces*. They are responsible for virtually all the molar cohesion of nonpolar polymers such as polyethylene, natural rubber, SBR, and butyl rubber. These forces act at a distance of approximately 0.4nm, and fall off rapidly, as the sixth power of the distance between atoms.② Consequently, molecules must be in close proximity for London forces to be effective. This helps to explain why a very flexible molecule such as natural rubber is a better adhesive than a moderately flexible molecule such as polystyrene. Low modulus, indicating freedom of rotation of submolecules that permits the adhesive to conform to the adherend, is advantageous to adhesion.

Interaction of permanent dipoles results in strong bonds, especially if the positive dipole is an H-atom. The *hydrogen bond*, typified by

$$\overset{-}{N}\!-\!\overset{+}{H}\cdots\overset{-}{O}\!=\!\overset{+}{C}$$

accounts for the excellent success with polar substrates of such diverse adhesives as starch and dextrin, polyvinyl alcohol, polyvinyl acetals, cellulose nitrate, phenolics, and epoxies. All of these adhesives contain phenolic or aliphatic hydroxyls. The carboxyl group incorporated in small percentage in many vinyl-type polymers, is an even more powerful aid to adhesion. Among the adherends utilizing H-bond adhesive are wood, paper, leather, glass, and metals.

——Skeist I, Miron J. Introduction to adhesives. In: Skeist I. ed, Handbook of Adhesives. 2nd ed. New York: Van Nostrand Reinhold Company, 1977. 11

Words and Expressions

adherend	[əd'hiərənd]	n.	被粘物
adhesive joint			黏合点
glue line			胶层
impervious	[im'pə:vjəs]	a.	不透水的，不渗透的
free energy			自由能
electrovalent	[i,lektrən'veilənt]	a.	电价的
covalent	[kəu'veilənt]	a.	共价的
heteropolar bond			极性键
homopolar bond			非极性键
weld	[weld]	n.; v.	焊接，熔焊
solder	['sɔ:ldə]	n.; v.	低温焊接
braze	[breiz]	n.; v.	钎接，焊接
Van der Waals' bonds			范德瓦耳斯键
dispersion force			色散力
molar cohesion			摩尔内聚力
dextrin	['dekstrin]	n.	糊精
contract	[kɔn'trækt]	v.	收缩

Phrases

be consistant with	与…一致	give rise to	造成，使发生
account for	说明		

Notes

① "This will occur provided the combining of adhesive…, and provided also that excessive strains are…" provided = if，引导条件状语从句。

② "…and fall off rapidly, as the sixth power of the distance between atoms." 译为："…以原子间距离的六次方迅速衰减。"

Exercises

1. *Complete the following paragraph with proper words*

    When two materials are bonded, the _____ is necessary. It must be _____ with the adherend. But adhesives often _____ as they harden. To avoid such shortcoming, we'd better choose _____ resins as adhesives, and they should be more _____ than the adherends. In addition, keeping the glue line as _____ as possible is important.

2. *Translate the following into English*

    极性键 非极性键 共价键 界面 表面 被粘物

Reading Materials

Acrylic Interpolymers Used as Pressure Sensitive Adhesives Crosslinked with Chelated Esters of Orthotitanic Acid

Pressure sensitive adhesive resins are widely employed in the form of filmlike coatings on a variety of superstrates to bond the latter to a normally nonadhering substrate. Once a bond has been made by applying a pressure-sensitive film between a substrate and a superstrate, the pressure-sensitive film may be subjected to substantial stress. Therefore, pressure-sensitive resins must exhibit a requisite high degree of adhesive strength as well as permanent tackiness.

The adhesive compositions of the process of *H. R. L. Gabriel*, *L. K. Post*, *B. M. Culbertson* and *C. M. Graham*; *U. S. Patent* 4 292 231; *September* 29 1981; *assigned to Ashland Oil*, *Inc*. comprise an adhesive polymer formed by the reaction of a metal alkoxide or a chelated metal alkoxide such as the chelated esters of orthotitanic acid with an interpolymer. The interpolymer comprises:

(1) at least 40 wt% of an ester of acrylic or methacrylic acid containing from 7 to 20 carbon atoms;

(2) 0.2 to 20 wt% of a vinyl type monomer which contains both a carboxylic acid and a carbamate functionality;

(3) Optionally up to 59.8 wt% of one or more additional copolymers which contain an ethylenically unsaturated linkage, such linkage being the only functional group within the monomer which will react with the metal alkoxide. Representatives of such monomers include vinyl esters of $C_3 \sim C_{10}$ alkanoic acids, ethyl and methyl esters of acrylic and methacrylic acids, acrylonitrile, styrene, vinyl chloride and the like.

The interpolymer formed from the monomers listed above are then crosslinked by metal alkoxides or chelated metal esters; especially preferred are chelated metal esters of orthotitanic acid.

The adhesives exhibit increased thumb appeal and adhesion promotion. The interpolymers thus prepared containing vinyl monomers with both carboxylic acid and carbamate functional groups exhibit a manageable viscosity at a reasonably high solids content and can be crosslinked at room temperature. Therefore, these compounds are ideal for use in preparation of pressure-sensitive adhesives.

Example: A reactor was charged with 668 parts 2-ethylhexyl acrylate, 48 parts ethyl acrylate, 174 parts vinyl acetate, 99.9 parts 2-melaoxyethyl carbamate and 512 parts ethyl acetate and slowly heated under N_2 to 85°C. At this point, 1 part of benzoyl peroxide was added. The mixture was maintained at 85 to 88°C for 45 minutes. Then 200 parts isopropanol, 200 parts methylene chloride and 1 part benzoyl peroxide were added and the mixture maintained at temperature 2 hours. Then 162 parts toluene, 260 parts ethyl acetate and 1 part benzoyl peroxide were added and maintained at 70 to 74°C for 1.5 hours. 1 parts benzoyl peroxide was then added and heated 1.25 hours. To the reactants were then added 316 parts isopropanol and 1 part benzoyl peroxide and heated at 70 to 74°C for 2.5 hours, at which point another 1 part benzoyl peroxide was added and the mixture heated at 70 to 73°C for 4.25 hours, upon cooling a conversion at 97.5% was obtained. Samples of the interpolymer solution were blended with Tyzor AA (a commercially available chelated ester of orthotitanic acid) and tested giving satisfactory results.

——Gutcho M. Adhesives Technology. Park Ridge: Noyes Data Publication, 1983.3

Words and Expressions

interpolymer	['intə'pɔlimə]	n.	共聚物
pressure sensitive adhesive			压敏胶
alkoxide	[æl'kɔksaid]	n.	醇盐，酚盐
chelate	['ki:leit]	v.	螯合
		n.	螯合物
orthotitanic acid			原钛酸
isopropanol	[aisəu'prəupənɔl]	n.	异丙醇

UNIT 33　Processing of Thermosets

　　Polymer processing is an engineering speciality concerned with the operations carried out on polymeric materials or systems to increase their utility.① These operations produce one or more of the following effects: chemical reaction, flow, or a permanent change in a physical property. Specially excluded are the chemical reactions involved in the manufacture of resins. Three factors are essential for any successful processing of polymers, namely materials, machinery, and process control.

　　Thermosetting polymers, as one of the most important materials, are more and more widely used as coatings, paints, foams, engineering plastics, reinforced plastics, and so on. With the thermoplastic polymer processing developing, the methods used to process thermosets are exploited one by one, especially the methods used to manufacture reinforced plastics.

　　A successful and profitable thermoset molding operation is dependent on the selection of the proper type of molding equipment, the plant layout, the method of molding, and molds that have been engineered properly and that are of superior construction. The methods of molding for thermoset processing mainly include compression, transfer, two-stage reciprocal screw transfer, injection, and extrusion (extrusion is a highly specialized and licensed process, production is quite limited). The basic methods of transfer molding are the pot and plunger method and the plunger transfer method.

　　Thermoset resins are by far the most widely used ones to produce reinforced plastics, largely because of their generally good properties, and relatively easy handling. Reinforced plastics are composites in which a resin is combined with a reinforcing agent to improve one or more properties of the plastic matrix. Although many types of reinforcements are used with plastics, the glass fibers predominate. Fibrous glass reinforcements are available in many forms: woven fabrics, nonwoven matting, bulk-chopped and milled fibers, and unidirectional rovings and yarns.②

　　Reinforced plastics can be fabricated by a wide variety of techniques to accomplish many different purposes. The techniques mainly include:

　　(1) Hand lay-up in which resin and glass fibers in the form of fabric, woven roving, or mat are simply placed in the mold manually;

　　(2) Contact molding in which the resin is in contact with the air and the wet lay-up normally hardens at room temperature;

　　(3) Vacuum-bag method in which cellophane or polyvinyl acetate film is placed over the lay-up, joints are sealed and a vacuum is created;

　　(4) Pressure-bag method in which a tailored bag, normally rubber sheeting, is placed against the lay-up;

　　(5) Autoclave, a modification of the pressure-bag method, in which the entire assembly is placed in a steam autoclave at 50 to 100 psi after lay-up;

　　(6) Spray-up in which chopped glass fiber roving (chopped strand) and catalyzed resin

are deposited simultaneously in a mold from special spraying equipment;

(7) Filament winding that combines continuous filamental reinforcement with a thermosetting resin binder;

(8) Matched metal die molding which is a compression molding process that uses a temperature-pressure-time cycle to polymerize a thermosetting resin combined with glass reinforcement;

(9) Bulk molding includes compression molding, transfer molding and injection molding with the putty-like mixture of thermosetting resin, filler, and chopped glass reinforcement;

(10) Sheeting molding in which the used material is a polymer resin mixture reinforced with glass strands and formed into a sheet that can be handled easily, cut to shape, and charged into a compression mold;

(11) Pultrusion which involves pulling the raw materials through shaping dies and curing operations;

(12) Continuous laminating in which fibrous glass fabricated mat is passed through resin dip and brought together between cellophane covering sheets.

By far the thermosetting resins have played a major role in manufacturing the advanced composites.

——Joel Frados. Plastics Engineering Handbook. New York:
the Society of the Plastics Industry, 1976. 462

Words and Expressions

thermoset	[ˈθəːməset]	n.	热固性（复数：热固性塑料）
plunger	[ˈplʌndʒə]	n.	活塞
matting	[ˈmætiŋ]	n.	垫子，席子，纺织品
roving	[ˈrəuviŋ]	n.	粗纱
yarn	[jɑːn]	n.	纱线
cellophane	[ˈseləfein]	n.	玻璃纸
autoclave	[ˈɔːtəkleiv]	n.	热压罐［釜］
strand	[ˈstrænd]	n.	线，绳
two-stage reciprocal screw transfer molding		n.	二阶双螺杆传递模塑
polyvinyl acetate		n.	聚乙酸乙烯酯
spray-up		n.	喷射成型
bulk molding compound (BMC)		n.	块状模塑料
sheeting molding compound (SMC)		n.	片状模塑料

Phrases

carried out the operation on　对…进行操作处理　　　　in the manufacture of　在…的生产中
be reinforced with　用…增强

Notes

① "Polymer processing is an engineering speciality concerned with the operations carried out on polymeric materials or systems to increase their utility." 译为："聚合物加工是一种工程专业，它与为了增加聚合物材料或体系的用途而进行的操作处理有关。" "carried out on polymeric materials or systems to increase their utility"这分词短语是用来修饰名词"operations。"

② "Fibrous glass reinforcements are available in many forms: woven fabrics, nonwoven matting, bulk-chopped and milled fibers, and unidirectional rovings and yarns." 译为: "玻璃纤维增强材料有许多形态, 诸如织物、无纺玻璃布、经切断和磨断的短纤维以及单向的粗纱和纱线。"

Exercises

1. *Put the following into Chinese*
   (1) sheeting molding
   (2) autoclave
   (3) two-stage reciprocal screw transfer molding
   (4) spray-up
   (5) bulk molding compound (BMC)
   (6) sheeting molding compound (SMC)

2. *Fill the blanks with proper words*
   Until now the most widely used polymer resins for producing reinforced plastics are _____ resins due to their good _____ and relatively easy _____. There are various forms of commercial fiber glass reinforcements, such as woven fabrics, nonwoven matting, bulk-chopped and milled fibers, and unidirectional rovings and yarns. The main techniques used to produce reinforced plastics include _____, _____, _____, _____, _____, and so on.

3. *Translate the third paragraph of the text into Chinese*

Reading Materials

A Composite Processing Technique——Resin Transfer Molding

In recent years, the complex geometry of composite parts has necessitated the development of new manufacturing processes to, eliminate conventional prepreg-lay-up. These processes include resin transfer molding (RTM) and filament winding. During resin transfer molding, the resin is injected into a mold containing dry fibrous preform. After the resin sets up, the composite structure is demolded and may be subjected to a free-standing post-cure.

The resin transfer molding process is gaining attention as an alternative to all of the hand craftsmanship of prepreg lay-up and the associated problems with that type of production. The RTM process produces low void content, complex shaped components with simple cure cycle. The designer has tight control of fiber volume and surface finish producing quality components with exact dimensional reproducibility. RTM has the following processing advantages over prepreg/ autoclave technique : control of all sides of part; no bagging labor or material costs; inclusive inserts, ribs, bosses, and cores in preform architecture; and high fiber loading. The higher the volume and complexity of the part, the more cost effective RTM becomes. So the major push for RTM in the aerospace sector is the tremendous cost saving that can be realized over the prepreg approach.

There are several requirements of a resin system to be used in an RTM process. These are : processibility, performance, compatibility, and low void formation. Processibility may be defined as ease of use under the conditions specified by the process. These are usually low viscosity, long working time, and short cure time. The lower the temperature required to achieve these characteristics, the better the resin system is generally accepted. Performance is entirely dependent on the end-use requirements for strength, stiffness, etc. The resin system must be compatible with the other materials utilized in the process to avoid corrosion or undesirable side reactions. Finally, no void formation is permitted once the resin system is inside the mold since there is no way to remove it. Therefore, resin systems should not generate gaseous by-products, emit volatile or outgas during processing or curing. A critical feature of the resin is that it has unique rheological properties. The process involves placement of dry reinforcement fiber in a closed mold which is subsequently injected with low viscosity resin, so the resin should maintain sufficiently low viscosity for sufficient time to allow good fiber wet-out. Resins with viscosities which are too high requiring higher rejection pressures can result in low quality parts due to poor fiber wet-out or fiber displacement. To be suitable for RTM, a resin must have sufficient time at a viscosity which can be pumped with conventional equipment (<1Pa · s, ideally, 0.5Pa · s) prior to reaching the mold. This "por-life" will vary with the application, but in some cases may need to be a full 8 hour work shift. The resin enters the mold which has been set at a higher temperature to reduce the resin viscosity and facilitate bulk flow and fiber wet-out. The resin is usually injected into the reinforcement fiber aided by vacuum and/or pressure.

These criteria pose a very real challenge to the resin supplier. The resins usually used for RTM have been room temperature or low temperature curable systems such as unsaturated polyester, vinyl ester, polyurethane, acrylates, epoxies, and bismaleimides. Especially epoxies and bismaleimides now are of much interest to the aerospace industry for structural components owing to their high performance.

——Frank C. Robertson. *British Polymer Journal*. 1988, 20: 417

Words and Expressions

prepreg	['priː'preg]	n.	预浸料
demold	[diː'məuld]	v.	脱模
polyurethane	[ˌpɔli'juəriθein]	n.	聚氨酯
acrylate	['ækrileit]	n.	丙烯酸酯
bismaleimide	[ˌbismæli'imid]	n.	双马来酰亚胺
lay-up		n.	铺层
post-cure		n.	后固化
wet-out		n.	浸渍

UNIT 34　Fillers for Polymers

Some films, fibers, and plastics are used as unfilled, polymers, but the strength and cost of most elastomers and plastic composites are dependent on the presence of appropriate fillers or reinforcements. Some rubber articles such as crepe rubber shoe soles, rubber bands, in ner tubes, and balloons are unfilled. However, the tread stock in pneumatic tires would not be serviceable without the addition of carbon black or amorphous silica. For example, addition of these fillers increases the tensile strength of SBR from 100 to 4000 psi. Likewise, most high-performance plastics are composites of polymers reinforced by fibrous glass.

Many naturally occurring functional materials, such as wood, bone, and feathers, are composites consisting of a continuous resinous phase and a discontinuous phase. The first synthetic plastics such as celluloid and Bakelite were also composites. Wood flour was used to reinforce the pioneer phenolic resins and is still used today for the same purpose.

According to the American Society for Testing and Materials standard ASTM D 883, a filler is a relatively inert material added to a plastics to modify its strength, permanence, working properties, or other qualities or to lower costs, while a reinforced plastics is one with some strength properties greatly superior to those of the base resin resulting from the presence of high-strength fillers embedded in the composition. According to ASTM, plastic laminates are the most common and strongest type of reinforced plastics.

According to one widely accepted definition, fillers are comminuted spherical or spheroidal solids. Glass beads which meet the requirements of this definition are used to reduce mold wear and to improve the quality of molded parts. The word extender, sometimes used for fillers, is not always appropriate since some fillers are more expensive than the resin.[①]

It is generally recognized that the segmental mobility of a polymer is reduced by the presence of a filler. Small particles with a diameter of less than 10 μm[②] increase the crosslinked density and active fillers increase the glass transition temperature (T_g) of the composite.

The high strength of composites is dependent on strong van der Waals interfacial forces. These forces are enhanced by the presence of polar functional groups on the polymer and by the treatment of filler surfaces with silanes, titanates, or other surface-active agents. Composites have been produced from almost every available polymer using almost every conceivable comminuted material as a filler.

Among the naturally occurring filler materials are cellulosics, such as wood four, α-cellulose, shell flour, starch, and proteinaceous fillers such as soybean residues. Approximately 40 000 tons of cellulosic fillers are used annually by the American polymer industry. Wood flour, which is produced by the attrition grinding of wood wastes, is used as a filler for phenolic resins, dark-colored urea resins, polyolefins, and PVC. Shell flour, which lacks the fibrous structure of wood flour, has been made by grinding walnut and peanut shells. It is used as a replacement for wood flour.

The addition of finely divided calcined alumina, corundum, or silicon carbide produces abrasive composites. However, alumina trihydrate (ATH), which has a Mohs' hardness of less than 3, serves as a flame-retardant filler in plastics. Zirconia, zirconium silicate, and iron oxide, which have specific gravities greater than 4.5, are used to produce plastics with controlled densities.

Calcium carbonate is widely used in paints, plastics, and elastomers. The volume relationship of calcium carbonate to resin or the pigment volume required to fill voids in the resin composite is called the pigment-volume-concentration (PIVC). The critical PIVC (CPIVC) is the minimum required to satisfy the resin demand.

Calcium carbonate is used at an annual rate of 7[3]* million tons as a filler in PVC, polyolefins, polyurethane foams, and in epoxy and phenolic resins.

Silica, which has a specific gravity of 2.6, is used as naturally occurring and synthetic amorphous silica, as well as in the form of large crystalline particulates such as sand and quartz.

Mica, which has a specific gravity of 2.8 and a Mohs' hardness value of 3, is a naturally occurring lamellar or platelike filler with an aspect ratio below 30. However, much higher aspect ratios are obtained by ultrasonic delamination.

Talc is a naturally occurring fibrous-like hydrated magnesium silicate with a Mohs' hardness of 1 and a specific gravity of 2.4. Since talc-filled polypropylene is much more resistant to heat than PP, it is used in automotive accessories subject to high temperatures. Over 0.4[3]** million tons of talc are used annually as fillers.

Nepheline syenite, a naturally occurring sodium potassium aluminum silicate filler, and wollastonite, an acicular calcium metasilicate filler, are used for the reinforcement of many plastics. Fibers from molten rock or slag (PMF) are also used for reinforcing polymers.

——Raymond B Seymour, Charles E Carraher. Jr. Polymer Chemistry. An Introduction. New York and Basel: Marcel Dekker, 1988. 360~365.

Words and Expressions

filler	['filə]	n.	填料，填充剂
reinforcement	[ˌriːinˈfɔːsment]	n.	增强剂
crepe	[kreip]	n.	(法语词) 绉沙，绉胶，绉片
pneumatic	[njuːˈmætik]	a.	气动的，气压的
		n.	气胎，有气胎的车辆
composite	[ˈkɔmpəzit]	n.	复合材料
laminate	[ˈlæmineit]	vt; n.	分层，层压；层压制品
comminute	[ˈkɔminjuːt]	vt.	粉碎，弄成粉末
segmental	[segˈmentəl]	n.	链段的
mobility	[məuˈbiliti]	n.	活动性，灵活性
silane	[ˈsilein]	n.	硅烷
titanate	[ˈtaitəneit]	n.	钛酸酯
alumina trihydrate		n.	氢氧化铝
zirconia	[zəːˈkəuniə]	n.	氧化锆
corundum	[kəˈrʌndəm]	n.	刚玉，金刚砂

calcine	['kælsin]	v.; n.	煅烧（矿），焙烧（矿、砂）
silica	['silikə]	n.	二氧化硅
mica	['maikə]	n.	云母
lamellar	[lə'melə]	a.	层［鳞、板、薄片、页片］状的
platelike	['pleit'laik]	a.	层［片］状的
aspect ratio		n.	长径［纵横］比，宽厚比
talc	[tælk]	n.	滑石
nepheline	['nefəlin]	n.	霞石
acicular	[ə'sikjulə]	a.	针状的
syenite	['saiinait]	n.	正长岩，黑花岗岩
wollastonite	['wɔːləstənait]	n.	硅灰石

Notes

① "The word extender, sometimes used for fillers..., are more expensive than the resin." 译为："人们常用增量剂一词来代替填料一词，但这种使用并非完全恰当，因为有些填料远比树脂昂贵。"（增量剂一词，以前用于填料，它表示增大树脂体积，但对其他性能贡献较小）。

② 此处原文为 10mm（用于填料的颗粒直径一般为微米级）。

③ * 原文为 700，** 原文为 40，根据1987年出版的《塑料填料与增强剂手册》一书，预测2000年填料用量为1500万吨，其中谈到 $CaCO_3$（calcium carbonate）消费量为900万吨，滑石（talc）消费量为180万吨）。

Exercises

1. *Translate the following paragraph into Chinese*

Many plastics are useful only when combined with reinforcing materials, such as particulates or fibrous solids. Amino resins and phenolic resins are almost always used in conjunction with fillers, which include wood flour, cellulose, powdered mica, asbestos, and glass fibers. These fillers greatly enhance dimensional stability, strength, abrasion resistance, and heat resistance. Glass fibers and woven fiberglass are widely used to reinforce unsaturated polyester resin and epoxy resins.

Strength properties of many thermoplastics such as nylon, polystyrene, acrylic resins, polycarbonates, acetal resins, and polyethylene are also greatly enhanced by the incorporation of glass fibers. A final and extremely important example is the reinforcement of rubber with various types of carbon blacks.

2. *Complete the following table by listing the fillers as many as you can*

Some fillers used in plastic industry

Spherical	
Plate-like	
Acicular	
Fibrous	

3. *Put the following into Chinese*

carbon black　　　　reinforcement　　　　continuous phase
discontinuous phase　　cross-linked density　　high performance plastics
surface active agent　　flame-retardant

4. *Translate the following into English*

增强剂　　　球形粒子　　　层状填料　　　复合材料　　　分散相
层压塑料　　炭黑　　　　　碳酸钙　　　　耐热性　　　　酚醛树脂

Reading Materials

Polymer Blending

In the preceding section we have seen how the properties of a given polymer may be modified by copolymerizing the initial monomer with an amount of another monomer sufficient to bring about the desired modifications in final properties. A reasonable question then arises: "Cannot the same desired end result be achieved by simply mixing one homopolymer with the right amount of the other homopolymer?" The answer to this question, generally speaking, is "no". It has long been known, and has been shown thermodynamically, that because of unfavorable free energy considerations polymers prefer to intertwine among themselves rather than among other polymers. Thus solutions of two different but electronically similar polymers dissolved in a common solvent, when mixed, eventually separate into distinct phases, each phase containing virtually all of only one polymer, and practically none of the other polymer. Separations of this type are also observed in the absence of solvent. Materials that separate so are said to be incompatible. Not only do chemical differences between polymers result in incompatibility but even structural differences can cause this effect.

Thus it has been shown that high-density polyethylene is not infinitely compatible with low-density polyethylene.

Because of their large size and somewhat limited mobility at normal temperatures, physical mixtures of polymers may very well stay together for very long times, and the thermodynamic tendency to separate might not be noticeable during the time of the experiment. Nevertheless, measurement of certain telltale properties distinguishes polymer blends from copolymers. The most critical of these are transition temperature measurements, either glass transition or melting point. Homopolymers or random copolymers show single transition unique for a single material, whereas blends or mixtures of polymers show a transition for each component present in the mixture.

With respect to properties such as tensile strength and other mechanical properties of blends, if the properties are determined by a single measurement during a short duration, the specimen appears to be a single material and the property is simply that of a weighted arithmetic average of the components of the blend. The above statement applies only to very well mixed and intimately blended polymer mixtures. Because amorphous polymers are most readily blended, ideal mixture behavior is most often found. in mixtures of amorphous polymers. Deviations from mechanical behavior in which blend properties are an average of homopolymer properties reflect poor mixing and most likely occur in crystalline polymer-amorphous polymer blends and crystalline polymer-crystalline polymer blends. Indeed, in some cases, it may be virtually impossible to get satisfactorily intimate mixing between crystalline polymers except by the use of a compatibilizing agent as discussed below.

This is not to imply that polymer blending is not a useful and important technique for preparing modified polymers. In the first place, in certain cases where molecular weights are fairly low, compatibilization can be realized. In the second place, if mixtures can be co-reacted, such as co-curing of two unsaturated rubbers or of melamine-formaldehyde with ureaformaldehyde resins, very useful materials result. Modification of alkyd resins with other polymers is a favorite procedure for varying the properties of surface coating resin. In this case, too, the polymer are co-cured so that the final product is actually a copolymer (held together by chemical bonds) rather than a simple physical mixture. In the third place, if the expected useful life of a fabricated article is relatively short and the article is not to be subjected to elevated temperatures where phase separation might occur rapidly, a simple mixture of polymers might be perfectly suitable in achieving a certain specific property in the article.

The fourth and the most common situation in which polymer blends are useful is when the blend is modified by an additional component called a compatibilizing agent Generally, these materials are to polymer blends what emulsifying agents are to oil-in-water or water in-oil emulsions. An emulsifying agent is usually thought of in terms of having head and tail portions, the head being soluble in one of the phases and the tail in the other phase Emulsifying agents are normally reasonably long molecules so that there is a reasonable distance between the head and the tail. If this analogy is carried through to polymer compatibilizers, they should likewise have a segment soluble in one polymer and a suitable distance away in the compatibilizer there should be a segment soluble in the other homopolymeric component. Of course, in polymer compatibilizers, the agent need not be composed of only two segments, but can be of many segments. It is not surprising, therefore, that the most successful compatibilizing agents are block and graft copolymers, where one segment is the same as one of the homopolymers, and the other segment is the same as the second.

Although most properties of stabilized polymer blends are the same as unstabilized blends, certain differences exist. For one, there is the presence of a third component which affects properties. More important, however, is the fact that in the presence of compatibiliz ers, homopolymers cannot migrate, and in the environment in which they find themselves, they tend to occupy the least volume. Thus they curl up or form tighter coils within each molecule, and within the whole of the material, the number of polymer-polymer entanglements is less than it would be in a homopolymeric mass, where chains are, extended to their most probable length and interchain entanglements are as frequent as intrachain entanglements. The most pronounced result of this state of affairs is that stabilized polymer blends show somewhat lower viscosities, that is, improved flow over what would be expected from the viscosities of each homopolymer. A compatibilized blend of two polymers having the same melt index number would be expected to have a higher melt index number (increased melt flow).

More often polymers are blended to improve properties other than melt flow properties. By far the most important and most common is the blending of rubbers with more brittle thermoplastics to improve toughness or impact strength.

The outstanding commercial products employing this technique are the impact polystyrenes, the ABS (acrylonitrile-butadiene-styrene) resins, and the impact methacrylate resins. Thus whereas polystyrene or polymethyl methacrylate homopolymers may have low impact strength, properly compatibilized blends of these materials with certain rubbers yield materials having impact strengths improved by as much as 10 to 20fold.

Although there are several techniques employed in the manufacture of these high impact materials, they have in common that as part of the production procedure block and graft copolymers are created. Thus impact polystyrene may be prepared by a vigorous co-mastication of polystyrene and a rubber such as a butadiene-styrene (SBR) copolymer. It may also be prepared by the polymerization of styrene monomer containing dissolved SBR.

Physically, a compatibilized polymer blend can be thought of as islands of homopolymer A in a sea of homopolymer B on one side of an inversion point, and islands of homopolymer B in a sea of homopolymer A on the other side of this inversion point, the inversion point being a function of the nature of the materials. The compatibilizing agents would then surround the islands and extend into the sea, preventing the islands from migrating and coalescing, in short, stabilizing a basically unstable situation.

With this model, it can be seen how each homopolymer maintains its identity, and how properties such as transitions are observed for each component; it is in this very fundamental way that properties of polymer blends differ from random copolymers having the same ratios of the same monomers.

——Herman S, Kaufman Joseph J Falcetta. Introduction to Polymer Science and Technology. New York: John wiley & Sons 1977, 124~127.

Words and Expressions

blend	[blend]	v.; n.	混合（物），共混（物），掺和（物），合金
incompatible	[inkəm'pætəbl]	a.	不相容的
melamine-formaldehyde resin		n.	三聚氰胺甲醛树脂，蜜胺树脂
urea-formaldehyde resin		n.	脲醛树脂
alkyd	['ælkid]	n.	醇酸（树脂）
emulsify	[i'mʌlsifai]	v.	使乳化
emulsion	[i'mʌlʃən]	n.	乳液，胶乳
compatibilizing agent		n.	增容剂
graft copolymer		n.	接枝共聚物
melt index number		n.	熔融指数
toughness	['tʌfnis]	n.	韧性，韧度
polymethyl methacrylate (PMMA)		n.	聚甲基丙烯酸甲酯，有机玻璃

APPENDIXES 聚合物的命名法

单　体（或原料）				聚　合　物	
分子结构	中文名称	英文名称	分子链结构	中文名称	英文名称
$CH_2=CH_2$	乙烯	ethylene	$—CH_2—CH_2—$	聚乙烯 （PE）	polyethylene
$CH_2=CH$ 　　$\|$ 　　CH_3	丙烯	propylene	$—CH_2—CH—$ 　　　　$\|$ 　　　　CH_3	聚丙烯 （PP）	polypropylene
$CH_3CH_2CH=CH_2$	1-丁烯	1-butene	$—CH_2—CH—$ 　　　　$\|$ 　　　　CH_2CH_3	聚-1-丁烯	poly-1-butene
CH_3 　$\|$ $CH_2=C$ 　$\|$ 　CH_3	异丁烯	isobutene	CH_3 　$\|$ $—CH_2—C—$ 　$\|$ 　CH_3	聚异丁烯	polyisobutene
$CH_2=CH—CH=CH_2$	1,4-丁二烯	butadiene	$—CH_2—CH=CH—CH_2—$	聚-1,4-丁二烯	poly-1,4-butadiene
CH_3 　$\|$ $CH_2=C—CH=CH_2$	异戊二烯	isoprene	CH_3 　$\|$ $—CH_2—C=CH—CH_2—$	聚-1,4-异戊二烯	poly-1,4-isoprene
Cl $\|$ $CH_2=C—CH=CH_2$	氯丁二烯	chloroprene	Cl $\|$ $—CH_2—C=CH—CH_2—$	聚-1,4-氯丁二烯	poly-1,4-chloroprene
$CH_2=CH$ 　　$\|$ 　　$\bigcirc$	苯乙烯	styrene (St)	$—CH_2—CH—$ 　　　　$\|$ 　　　　$\bigcirc$	聚苯乙烯 （PSt）	polystyrene

续表

单体(或原料)			聚合物		
分子结构	中文名称	英文名称	分子链结构	中文名称	英文名称
$CH_2=C(CH_3)-C_6H_5$	α-甲基苯乙烯	α-methylstyrene	—CH$_2$—C(CH$_3$)(C$_6$H$_5$)—	聚-α-甲基苯乙烯	poly-α-methylstyrene
$CH_2=CHCl$	氯乙烯	vinylchloride (VC)	—CH$_2$—CHCl—	聚氯乙烯 (PVC)	polyvinylchloride
$CH_2=CCl_2$	偏二氯乙烯	vinylidenechloride	—CH$_2$—CCl$_2$—	聚偏二氯乙烯 (PVDC)	polyvinylidenechloride
$CH_2=CHF$	氟乙烯	vinylfluoride	—CH$_2$—CHF—	聚氟乙烯 (PVF)	polyvinylfluoride
$CF_2=CF_2$	四氟乙烯	tetrafluoroethylene	—CF$_2$—CF$_2$—	聚四氟乙烯 (PTFE)	polytetrafluoroethylene
$CF_2=CFCl$	三氟氯乙烯	trifluorochloroethylene	—CF$_2$—CFCl—	聚三氟氯乙烯	polytrifluorochloroethylene
$CH_2=CH-O-CO-CH_3$	乙酸乙烯酯	vinylacetate (VAC)	—CH$_2$—CH(OCOCH$_3$)—	聚乙酸乙烯酯 (PVAC)	polyvinylacetate
—CH$_2$—CH(OCOCH$_3$)—	聚乙酸乙烯酯	polyvinylacetate	—CH$_2$—CH(OH)—	聚乙烯醇 (PVA)	polyvinylalcohol
—CH$_2$—CH(OH)—CH$_2$—CH(OH)— + RCHO	聚乙烯醇醛	polyvinylalcohol aldehyde	—CH$_2$—CH—CH$_2$—CH— with O—CH(R)—O bridge	聚乙烯醇缩醛	polyvinylacetal

APPENDIXES 聚合物的命名法

续表

单 体（或原料）			聚 合 物		
分子结构	中文名称	英文名称	分子链结构	中文名称	英文名称

—CH$_2$—CH
       |
       OH
H—C═O
     |
     H
	聚乙烯醇甲醛	polyvinylalcohol formal dehyde	—CH$_2$—CH—CH$_2$—CH—
			          | |
			          O—CH$_2$—O
				聚乙烯醇缩甲醛	polyvinylformal

CH$_2$═CH
      |
      CN
	丙烯腈	acrylonitrile
		(AN)
			—CH$_2$—CH—
			        |
			        CN
				聚丙烯腈
				(PAN)
					polyacrylonitrile

CH$_3$
      |
CH$_2$═C
      |
      CN
	α-甲基丙烯腈	α-methylacrylonitrile	CH$_3$
			        |
			—CH$_2$—C—
			        |
			        CN
				聚-α-甲基丙烯腈	poly-α-methylacrylonitrile

CH$_2$═CH
      |
      C═O
      |
      NH$_2$
	丙烯酰胺	acrylamide
		(AM)
			—CH$_2$—CH—
			        |
			        C═O
			        |
			        NH$_2$
				聚丙烯酰胺
				(PAM)
					polyacrylamide

CH$_3$
      |
CH$_2$═C
      |
      C═O
      |
      NH$_2$
	α-甲基丙烯酰胺	α-methylacrylamide	CH$_3$
			        |
			—CH$_2$—C—
			        |
			        C═O
			        |
			        NH$_2$
				聚-α-甲基丙烯酰胺	poly-α-methylacryl-amide

CH$_2$═CH
      |
      COOH
	丙烯酸	acrylic acid
		(AA)
			—CH$_2$—CH—
			        |
			        COOH
				聚丙烯酸
				(PAA)
					polyacrylic acid

CH$_2$═CH
      |
      COOR
	丙烯酸酯	acrylate	—CH$_2$—CH—
			        |
			        COOR
				聚丙烯酸酯	polyacrylate

CH$_2$═CH
      |
      COOCH$_3$
	丙烯酸甲酯	methylacrylate	—CH$_2$—CH—
			        |
			        COOCH$_3$
				聚丙烯酸甲酯	polymethylacrylate

续表

单 体（或原料）			聚 合 物		
分子结构	中文名称	英文名称	分子链结构	中文名称	英文名称
$\begin{array}{c}\text{CN}\\ \text{CH}_2\!=\!\text{C}\\ \text{COOR}\end{array}$	α-氰基丙烯酸酯	α-cyanoacrylate	$\begin{array}{c}\text{CN}\\ -\text{CH}_2-\text{C}-\\ \text{COOR}\end{array}$	聚-α-氰基丙烯酸酯	poly-α-cyanoacrylate
$\begin{array}{c}\text{CH}_3\\ \text{CH}_2\!=\!\text{C}\\ \text{COOH}\end{array}$	甲基丙烯酸	methacrylic acid (MAA)	$\begin{array}{c}\text{CH}_3\\ -\text{CH}_2-\text{C}-\\ \text{COOH}\end{array}$	聚甲基丙烯酸 (PMAA)	polymethacrylic acid
$\begin{array}{c}\text{CH}_3\\ \text{CH}_2\!=\!\text{C}\\ \text{COOR}\end{array}$	甲基丙烯酸酯	methacrylate	$\begin{array}{c}\text{CH}_3\\ -\text{CH}_2-\text{C}-\\ \text{COOR}\end{array}$	聚甲基丙烯酸酯	polymethacrylate
$\begin{array}{c}\text{CH}_3\\ \text{CH}_2\!=\!\text{C}\\ \text{COOCH}_3\end{array}$	甲基丙烯酸甲酯	methylmethacrylate (MMA)	$\begin{array}{c}\text{CH}_3\\ -\text{CH}_2-\text{C}-\\ \text{COOCH}_3\end{array}$	聚甲基丙烯酸甲酯 (PMMA)	polymethylmethacrylate
$\begin{array}{c}\text{CH}_2\!=\!\text{CH}\\ \text{O}\\ \text{R}\end{array}$	乙烯基醚	vinylether	$\begin{array}{c}-\text{CH}_2-\text{CH}-\\ \text{O}\\ \text{R}\end{array}$	聚乙烯基醚	polyvinylether
$\begin{array}{c}\text{CH}_2\!=\!\text{CH}\\ \text{O}\\ \text{CH}_3\end{array}$	乙烯基甲基醚	vinylmethylether	$\begin{array}{c}-\text{CH}_2-\text{CH}-\\ \text{O}\\ \text{CH}_3\end{array}$	聚乙烯基甲基醚	polyvinylmethylether
顺丁烯二酸酐环结构	顺丁烯二酸酐（马来酸酐）	maleic anhydride	聚顺丁烯二酸酐环结构	聚顺丁烯二酸酐（聚马来酸酐）	polymaleic anhydride

续表

单体（或原料）			聚合物		
分子结构	中文名称	英文名称	分子链结构	中文名称	英文名称
CH₂=CH–N(C=O)(CH₂)₃ (环)	乙烯基吡咯烷酮	vinylpyrrolidone	–CH₂–CH–N(C=O)(CH₂)₃	聚乙烯基吡咯烷酮 (PVP)	polyvinylpyrrolidone
CH₂=CH–C(=O)–CH₃	乙烯基甲基酮	vinylmethylketone	–CH₂–CH–C(=O)–CH₃	聚乙烯基甲基酮	polyvinylmethylketone
CH₂=CH–N(carbazole)	乙烯咔唑	vinylcarbazole	–CH₂–CH–N(carbazole)	聚乙烯咔唑	polyvinylcarbazole
CH₂=CH–(2-pyridyl)	2-乙烯基吡啶	2-vinylpyridine	–CH₂–CH–(2-pyridyl)	聚-2-乙烯基吡啶	poly-2-vinylpyridine
HO–R–OH HO–C(=O)–R'–C(=O)–OH HO–R"–C(=O)–OH	二元醇 二元酸 羟基酸	glycol(diol) dibasic acid hydroxy acid	–R–O–C(=O)–R'–C(=O)– –R"–C(=O)–O–	聚酯	polyester
HO–CH₂CH₂–OH HO–C(=O)–C₆H₄–C(=O)–OH	乙二醇 对苯二甲酸	ethylene glycol tere-phthalic acid	–CH₂CH₂–O–C(=O)–C₆H₄–C(=O)–	聚对苯二甲酸乙二醇酯（涤纶；PET）	polyethylene tere-phthalate(Dacron)

续表

单体（或原料）			聚合物		
分子结构	中文名称	英文名称	分子链结构	中文名称	英文名称
$H_2N-R-NH_2$ $HO-\overset{O}{\underset{\|}{C}}-R'-\overset{O}{\underset{\|}{C}}-OH$	二元胺 二元酸	diamine dibasic acid	$-R-NH-\overset{O}{\underset{\|}{C}}-R'-\overset{O}{\underset{\|}{C}}-N-$	聚酰胺 （尼龙；耐纶）	polyamide (Nylon)
$H_2N-(CH_2)_6-NH_2$ $HO-\overset{O}{\underset{\|}{C}}-(CH_2)_4-\overset{O}{\underset{\|}{C}}-OH$	己二胺 己二酸	hexamethylenediamine adipic acid	$-(CH_2)_6-NH-\overset{O}{\underset{\|}{C}}-(CH_2)_4-\overset{O}{\underset{\|}{C}}-NH-$	聚己二酸己二酰胺 （尼龙 66；耐纶 66）	polyhexamethylene adipamide (66Nylon)
$(CH_2)_5\overset{\overset{\displaystyle C=O}{\|}}{\underset{\displaystyle NH}{\|}}$	己内酰胺	caprolactam	$-NH-\overset{O}{\underset{\|}{C}}-(CH_2)_5-$	聚己内酰胺 （尼龙 6；锦纶）	polycaprolactam
$HO-R-O-H$	二元醇	glycol(diol)	$-R-O-$	聚醚	polyether
$\overset{CH_2=O}{\underset{H}{\|}}$	甲醛	formaldehyde	$-CH_2-O-$	聚甲醛	polyoxymethylene (polyformaldehyde)
$\overset{\displaystyle CH_2-CH_2}{\underset{\displaystyle O}{\diagdown\diagup}}$	环氧乙烷 （氧化乙烯）	ethylene oxide	$-CH_2CH_2-O-$	聚环氧乙烷（聚氧乙烯；聚乙二醇）	polyethylene oxide (polyoxyethyleneglycol)
$\overset{\displaystyle CH_2-CH-CH_3}{\underset{\displaystyle O}{\diagdown\diagup}}$	环氧丙烷 （氧化丙烯）	propylene oxide	$-CH-CH_2-$ $\underset{\|}{CH_3}$	聚环氧丙烷 （聚氧化丙烯）	polypropylene oxide
$\overset{\displaystyle CH_2-CH-CH_2}{\underset{\displaystyle Cl}{\diagdown\diagup\atop O}}$	环氧氯丙烷	epichlorohydrin			
$HO-\bigcirc-\underset{\underset{CH_3}{\|}}{\overset{\overset{CH_3}{\|}}{C}}-\bigcirc-OH$	双酚 A	bisphenol A	环氧树脂结构	环氧树脂	epoxyresin
$\bigcirc-OH$	苯酚	phenol	酚醛树脂结构	酚醛树脂	phenolic resin
$\overset{O}{\underset{H}{\|}}H-C=O$	甲醛	formaldehyde			

续表

单体（或原料）			聚合物		
分子结构	中文名称	英文名称	分子链结构	中文名称	英文名称
$H_2N-C-NH_2$ (O) / $H-C-H$ (O)	尿素 / 甲醛	urea / formaldehyde	$-CH_2-NH-C-NH-$ $\|$ $-CH_2-N-C-N-$ (O, CH$_2$, O)	脲醛树脂	urea-formaldehyde resin
三聚氰胺结构 / $H-C-H$ (O)	三聚氰胺 / 甲醛	melamine / formaldehyde	(三嗪环交联结构)	三聚氰胺甲醛树脂（蜜胺树脂）	melamine formaldehyde resin
OCN—R—NCO / HO—R′—OH	二异氰酸酯 / 二元醇	diisocyanate / diol	$-R-O-C-NH-R'-NH-C-O-$ (O, O)	聚氨酯 (PU)	polyurethane
CH_3—〇H—CH_3	2,6-二甲基苯酚	2,6-dimethylphenol	(对甲基苯氧撑结构)	聚-2,6-二甲基苯酚（聚二甲基苯氧撑）	poly-2,6-phenylene oxide
均苯四酸二酐结构 / H_2N—〇—O—〇—NH_2	均苯四酸二酐 / 4,4′-二氨基二苯醚	pyromellitic dianhydride / 4,4′-daminodiphenylether	(聚酰亚胺结构)	聚酰亚胺	polyimide

续表

单体（或原料）			聚合物		
分子结构	中文名称	英文名称	分子链结构	中文名称	英文名称
（双酚A结构）； （4,4'-二氯二苯砜结构）	双酚 A； 4,4'-二氯二苯砜	bisphenol A; 4,4'-dichlorodiphenol sulfone	（聚砜结构式）	聚砜	polysulfone
（双酚A结构）； Cl—C(=O)—Cl	双酚 A； 光气	bisphenol A; phosgene	（聚碳酸酯结构式）	聚碳酸酯	polycarbonate
（二氯甲基丁氧烷结构）	二氯甲基丁氧烷	bischloromethyloxa-cyclobutane	（聚二氯甲基丁氧烷结构）	聚二氯甲基丁氧烷（氯化聚醚）	polydichloromethyoxa-cyclobutane(chlorinated polyether)
Br—⌬—SCu	对溴硫代苯酚亚铜	cuprous p-bromo-thiophenoxide	—⌬—S—	聚苯硫醚	polyphenylene sulfide
（3,3'-二氨基联苯胺结构）； （间苯二甲酸结构）	3,3'-二氨基联苯胺； 间苯二甲酸	3,3'-diaminobenzidin; isophthalic acid	（聚苯并咪唑结构）	聚苯并咪唑	polybenzimidazole

续表

单体（或原料）			聚合物		
分子结构	中文名称	英文名称	分子链结构	中文名称	英文名称
(structure)	均苯四酸二酐	pyromellitic dianhydride	(structure)	聚苯并咪唑吡咯烷酮	polyimidazopyrrolone
(structure)	均苯四胺	pyromellitoamine			
$CH_3-Si(Cl)_2-CH_3$	二甲基二氯硅烷	dimethyldichlorosilane	$-Si(CH_3)_2-O-$	硅树脂	polydimethylsiloxane (silicone)
$Cl-CH_2CH_2-Cl$ Na_2S_4	1,2-二氯乙烷 四硫化钠	1,2-dichloroethane sodiumtetrasulfide	$-CH_2-CH_2-S-S-$	聚硫橡胶	polysulfide rubber
$Br-CH_2-C(O)-CH_2-Br$ $H_2N-C(S)-R-C(O)-NH_2$	二溴丙酮 二硫酰胺	dibromoketone dithioamide	(thiazole ring structure)	聚噻唑	polythiazole
(glucose structure)	葡萄糖	glucose	(cellulose structure)	纤维素	cellulose